LAND UTILIZATION

By the same Author

* Frontiers in Migration Analysis (ed.), 1981, Rs. 160
* Introduction to Rural Settlements, 1979, Rs. 90
* Introductory Methods in Population Analysis, 1989, Rs. 200
* Patterns of Regional Geography: An International Perspective (ed.), (in 3 vols.) 1990, Rs. 1000 (per set)
* Planned Development of Rural Settlements, 1981, Rs. 150
* Recent Trends and Concepts in Geography (eds.) (in 3 vols.) 1980, Rs. 400 (per set)
* Statistics for Geographers and Social Scientists, 1982, Rs. 175
* Systems of Rural Settlements in Developing Countries, 1989, Rs. 540
* Urbanization and Regional Development (ed.), 1982, Rs. 170

LAND UTILIZATION

Theory and Practice

R.B. MANDAL

CONCEPT PUBLISHING COMPANY, NEW DELHI 110059

First Published 1982
2nd Reprint 1990

Published and Printed by
Ashok Kumar Mittal
Concept Publishing Company
A/15-16, Commercial Block, Mohan Garden
NEW DELHI 110059 (India)

Lasertypeset by
Microtech Advance Printing Systems (P) Ltd.
H-13, Bali Nagar
NEW DELHI - 110015

Preface

LAND IS the basic resource of human society. Its utilization shows a reciprocal relationship between the prevailing ecological conditions of a region and man. The term 'land use' is employed for varied surfacial and dynamic uses of the land and soil surveys e.g., land under cultivation, pasture, barren, orchard, fallow, waste, culturable waste, settlement, forests, water bodies, etc. on varied terrain conditions and soil types.

The study of land use is of prime importance for the economic development, resource planning and cultural advancement of people. The potentiality of a region depends upon the intensive or extensive use of the land. The intensive use of land bears population concentration, economic prosperity through better agricultural production, human establishments, industrial location, communication and transportation lines, while extensive use of the land bears sparse population, dispersed settlements, solitary living, agricultural economy and poor transport network. However, it is only through the systematic utilization of the land that economic and cultural development can be made possible. Thus one cannot think of progress without a proper utilization of land. Hence, the study of land use is of immense value in any scheme of regional development and planning.

The author is extremely grateful to Dr P. Dayal M.A. (Alig.), Ph.D. (London), Vice-Chancellor, Magadh University for his kind inspiration and ready guidance at all levels. I am highly thankful to Dr A. Sharan, Reader in Geography, Bihar University and Dr R.N.P. Sinha, Reader in Geography, Patna University for their genuine interest in land use studies. I am indebted to Dr L.N. Ram, Reader in Geography, Patna University under whose guidance and supervision I have been able to complete the whole work. I am also thankful to Dr G.N. Singh, Lecturer in Geography and P.N. Mandal,

Lecturer in Philosophy, R.D. and D.J. College, Monghyr, whose cooperation, inspiration and guidance have been most fruitful in the completion of this study. I invite constructive criticisms and scholarly suggestions from my learned readers for further improvement of this work in the next edition.

RAM BAHADUR MANDAL

Contents

List of Figures

List of Tables

1

Introduction

LAND USE is the surface utilization of all developed and vacant land on a specific point, at a given time and space. This "leads one back to the village farm and the farmer, to the fields, gardens, pastures, fallow land, forests and to the isolated farmstead" (Freeman, 1968)[1] as geography deals with the spatial relationship between these aspects and planning. This is because land use changes to meet the variable demands of the land by the society in its new ways and conditions of life. The demand for new uses of land may be inspired by a technological change, or by a change in the size, composition and requirements of a community. Some changes are short-lived, whereas others represent a more constant demand (Jackson, 1963).[2] A clear understanding of these dynamic qualities in land use will emerge from a historical survey designed to reveal the successive development of inherent characteristics of land.

In this way, land utilization is the use made of the land by man, as surveyed and mapped in a series of recognised categories. The primary uses of land are for crops, forest, pasture, mining, transportation, gardening, residential, recreational, industrial, commercial and uncultivable waste, barren and fallow land, etc. It is not normally possible to use land for two or more purposes simultaneously though sometimes even this is possible i.e., the pastured woodland. In most cases the proportion of waste land is quite large.

The concept of land use planning has been recently introduced in land utilization studies which means the formulation and administration of land policies aimed at the employment of land resources and the uses for which they are socially, politically and economically best suited.

The idea of a land use hierarchy varies with the production and consumption factors. The production factors are land, transportation facilities and the stage of technological advancement. The consumption factor include the number of people, consumption of goods per person and the gross exports. Any shift in the consumption of commodities may change the consumption of other commodities. The growth of population may change the forest and pasture land into cropland, including residential and industrial land utilization survey made uptill now were mostly concerned with the smaller areas of rural and urban sectors. This involves a description of the use of the land, trend in utilization and the classification of the land based on physical and economic factors. The data of land tenure, farm area and non-farm area are collected in order to know the productivity of several classes of land. Whether the land is used of crops, pasture and forest should be studied so that the comparison of yields, values and the costs of agricultural commodities can be made. This leads to the study of systems of farming on various categories of land, and economic prosperity of the region as well. With the help of land utilization survey, probable change in the use of land can be estimated in close connection with the institutional, social and public expenditure in the area. But the regional and national treatment of land use studies receive area attention of the scientist. Thus the study of land utilization has both economic, geographic and demographic dimensions. The geographic aspect consists largely of a survey of the temperature, humidity, topography and soil conditions which influence the utilization of land for crops, pasture or forest. The demographic aspect considers the studies of population distribution, composition, characteristics and trends not only in the area being surveyed, but in the whole country.

MEANING OF LAND USE

The difference between land use and land utilization is important. Land use is the use actually made of any parcel of land, House, apartments and industrial location are land use categories, whereas the term residential, industrial and agricultural refers to a system of land utilization implying roads, neighbourhood retail and service activities as well as location of industries, and the carrying of agricultural pursuits. In a rural area, tree crop or row crop would identify land use, whereas orcharding, truck farming and grazing indicate a system of land utilization.

Land is the basic resource of human society. Its utilization shows a reciprocal relationship between the prevailing ecological conditions of a particular region and man. The term 'land utilization' is also used for varied utilization of land and soil surveys, e.g., land under cultivation, pasture barren, orchard, fallows, waste, culturable waste, settlements, forests, water-bodies, etc. According to J.L. Buck "land utilization is the satisfaction, which the farm population derives from the type of agriculture developed, the provision for future production and the contribution to national needs" (Quoted, 1951).[3] While the definition given by Salter is as follows: "Land utilization research can be described as dealing with problem situations in which people in a given locality are in the process of transformation from activities with certain land requirements to activities with different land requirements" (Kelso, 1962).[4] In this sense land utilization involves an examination of the natural factors affecting both the harnessed and the potential productivity of the land in a changed situation of the locality and its requirements. These factors are the land, temperature, rainfall and soil, which in a configuration together constitute the physical background of agriculture and determine the limits of both the culturability and productivitv of the land.

Land utilization mainly deals with the problems related to the society and the region as a whole, rather than a private farmer. Land use is mainly related to the optimum use of the limited land between the alternative major types of land use. In rural areas, the major types of land use is as follows: (1) Agricultural land (a) non-irrigated lands, (b) irrigated lands, (c) dry farming areas, (d) grazing areas; (2) Village orchards or forest lands; (3) Forest land (a) forest covered, (b) forest reclaimed land use, (c) culturable or recreational land use.

Land utilization is also related to "Conversion of land from one major use to another general use" (Nanavati, 1957).[5] After reclamation of forest land, a question arises as to how the land should be utilized. The rotation of crops and their combination are after all minor problems of land use study. This is because these aspects depend upon personal experience and intuition of the farmer who decides which crops should be grown in rotation.

There is an intimate relationship between land economics and land utilization. The efficient use of land depends on the capacity of the man to utilize the land and to manage it. It also depends upon the systems of farming, systems of land tenures and size of land holdings. Whereas, the production efficiency and level of production depend upon the institutional framework and the production function carried on by the farmer.

The man-land relationship can be expressed in three different aspects. First, the land and the individual person who uses it; second, the man and his influence on the use of the land as a means of production. In this case institutional infrastructure should be studied for improved land use. The third relationship between land and man can be expressed in terms of man as a social being and the land as an inexhaustible resource.

Importance of the subject : For human existence, within certain biotic, ecological and economic conditions the utilization of land is of prime importance. It involves a relationship that exist between the society on the one hand, and cultural advancement, resource planning and carrying capacity of the land, on the other. The intensive use of land depends upon population concentration, economic prosperity through better agricultural production, human establishments, industrial locations, communication and transport lines, while extensive use of the land is related to sparse population, dispersed settlements, the absence of communication lines and the crude forms of transport. However, only the systematic utilization of land can be able to promote economic and cultural advancement. If there is no utilization of land, one cannot think of any progress. Thus the study of land utilization is of immense value in tracing out the past use of land and its future trend. Only through the study of the past land utilization, one can be able to predict its future use and evolve land use planning of a particular region. The suggestion for better use of land and the improvement of its present capability necessitate mapping and interpreting the utilization of land. The changing population and the economic, the biological and the ecological problems are so alarming that the conservation and the best utilization of land becomes a necessity.

Scope and objectives of land use studies : Land use is the latest ramification of the fast growing tree of economic geography. Geographers can present a clear picture of the potentials of land use, conducive to fruitful planning for a massive agricultural turnover. A plan for the utmost utilization of land asset should essentially include the requirements of land for location of industries, sites for houses, schools, public buildings, railways, roads, canals and airports etc. Sites for defence establishments or big government offices, parks, playing grounds, clubs and cinema houses all are included as items of any land utilization survey. As one use of land excludes another, it is necessary to assess fully the potentialities of every particular type of land in respect of all possible uses including the utilities, industrial, residential and the other. But utilization of land for items other than farming should not be blind foldedly pursued as it would affect the farming needs. The consumption of better agricultural land for purposes of urbanization is, thus, a great danger to the farmers; because the land withdrawn from agricultural use is either never returned to the farmer or even is returned, it is so done only after a long lapse of time (Stapledon, 1940).[6]

The scope of the present study confines itself to the essential ingredients of land use, i.e., factors principles, approaches, land use classification, agricultural efficiency, crop combination, land capability, conservation of land and land use planning. Factors of land use encompasses physical, economic and institutional aspects. Principles of land use represents superiority of one use of the land over the other which is determined by the social needs and cultural advancement of the people. The study of the approaches of land use includes Malthusian and Marxian approaches, besides, it deals with Von Thunen model of land use intensity. Further, Polar and anti-polar concepts of land use are also discussed here which are modified forms of Von Thunen model.

The techniques of land use study include field investigation, preparation of field maps, a short scheme for village survey, designs of sampling and the presentation of formulas investigated by learned scholars in diverse fields of land use study. The scope of land utilization was the development and the application of land classification techniques. The extent was to map geographically the patterns of physical and biological characteristics of the environment. Generally, they were descriptive inventories of resources described cartographically and how the inventories might be used in economic decision-making was quite unclear (Kelso, 1962).[7] Land use classification is given here in a broader dimension 'covering the world iand use and the land use studies of the U.K., India, the U.S.A., Japan, the U.S.S.R. which are followed by a case study from Bihar.

The discussion on agricultural efficiency mainly covers food production efficiency among different crops as a case study of Bihar. Further more, a new method of land capability classification has been given which measures the potential of land after considering the effects of positive and negative variables on land. The measurement of land resource is assessed here in terms of 'food production efficiency per unit area' and its conversion into calories, because "it may not be fair to equate a pound of wheat with a pound of millet as their caloric yields are vastly different" (Shafi, 1969).[8] So far as the crop combination is concerned, several methods have been discussed here along with a new method which has the advantage of accompanying dominant crops of the region discarding diversified nature of smaller amount of crops grown. The chapter on land reforms deals with special reference to India, as this is prerequisite rather than necessary for ushering an era of social equality and economic growth of any region. At present most of our productive land is concentrated in a few hands and the population is rapidly increasing, specially that of the low income group this has created a very tense situation among the land owners and the landless labourers. Thus the measure of land reforms should be emphatically considered for any scheme of land tenurial modification. In the end land use planning is discussed keeping in view the problems which the society is facing in the field of land utilization. For better land use a provision has been made for farm management which is concerned with the full employment for the society as a whole as India is a populous country where the mechanisation of land holding should not be adhered to. With the increase in the scope of land use, geographers study the use of land for regional analysis, while economists do it for resource appraisal. The study of soil series (1927)[9] and its conservation also comes under land utilization. This helps in checking the erosion and leaching of soil to maintain its productivity. Kellog's (*Readings in Land Utilization* 1957)[10] has studied the problem of conservation of agricultural land in his paper entitled 'Conflicting Doctrines About Soils'. He enlists seven schools of thought to understand the soil complex. First of all, he mentions 'The Balance Sheet Theory' which takes into account the plant nutrients in soil. In terms of this theory, the nutrients are stored in a storage bin which can be taken out of soil as per requirement and be put back into the soil through manures and fertilizers, etc. Secondly, he mentions the 'erosion menace school' which is concerned about the erosion of soils, the third mentioned is 'the organic school' which stresses the organic factors in soil management and several others. There is a certain amount of truth in all these but for correct understanding of the problem precise studies need to be carried out.

Thus, the determination of soil fertility is a complex phenomenon. It cannot be explained merely in terms of the physical and chemical properties of the soil. Factors such as aeration, structure, texture and the age together with the methods of management of the soil also need to be taken into account. The type of soil is an important factor determining its suitability for growing a certain variety of crop. Steady maintenance of the productive power of the soil is another important aspect in land utilization. Developed nations like the U.S.A. and the U.S.S.R. have completed extensive soil surveys and mapped out all types of soil profiles. In the U.S.A. the State Planning Boards have been started marking the culmination of land utilization surveys conceived on the broadest possible basis. China has also completed a land utilization survey. The U.K. and Denmark, having industrial and agricultural economic base, brought out the land utilization survey on the top of the list of countries in the world.

The objectives of land use study also goes to prevent

soil erosion and facilitate economic use of the land. It is, thus, necessary to classify the land according to the productivity, workability and conservability. Proper management of grazing land is also essential for better livestock raising. According to Forsling the deteriorating condition of livestock raising is mainly determined by the factors, i.e., (1) close and unsupervised grazing by animals, (2) drought, (3) improper land tenure system and nomadic grazing, (4) indiscriminate burning of grazing land, (5) invasion of unpalatable species, and (6) losses due to flood. He suggested improved techniques of grazing, land management and intensive study of social and economic parameters preventing such wide use.

Afforestation is the best use of land to control the soil erosion. Agricultural operation may be economical at the cost of reduced timber supply and firewood, but this will pave the way for soil erosion. Thus the best study of land is to classify it according to productivity for preventing its improper use.

The land use study should be made in close connection with the water resource management, flood protection scheme and irrigation utilities. Such study should include : (a) Survey of soils and classification of land, (b) Investigation of surface and underground water resource for irrigation, and (c) Re-examination of soils after the beginning of irrigation from time to time.

These facts help in determining the changes in the root zone of subsoil drainage or the accumulation of soluble salts on the surface layers.

In India the land use survey is mostly carried on the pattern of Britain where L. Dudley Stamp played a vital role in 1930's in preparing the land use map of entire U.K. This survey had highly contributed in enriching the food production of U.K. and reorienting the agriculture on more scientific basis (Sharan, 1969)[11]

On the other hand, this survey in India has not been taken up in a comprehensive way. This may help in better land resource planning and management. It has yet to receive the blessings of the Royal Commission on agricultue. The population growth is high during recent decades but the per capita income is exorbitantly low, which requires land utilization survey with a view to extracting the utmost possible produce from the land. Such survey would bring to light the unharnessed potentialities of the cultivated and uncultivated land and would also boost up our efforts in the direction of preventing soil decay by erosion and in maintaining the levels of productivity. Problems of waterlogging, land reclamation, drainage, irrigation and other technical aspects can be profitably tackled by land use survey alone.

Keeping in view these ideas, the State of Bihar is in dire need of land use survey and farm management for agricultural development. By arresting population growth, checking the fragmentation of land holdings, and with the use of modern techniques of farming, provision of irrigation facilities and means of quick transport, we can achieve land use for rural development. The adoption of such practices require careful mapping and thorough planning of the present land use. Recently, the Central and the State Planning Boards have taken up these issues in a right earnest but these seem to be expensive. Moreover, there is a lack of due consideration of the various complex facets of land use as envisaged by the eminent earth scientists, whose attempts have been praiseworthy even if their works are mainly academic in nature.

Early land use studies : The importance of land utilization is being felt with the passage of time. Such studies are important from two aspects: (a) academic, as a specified field of training and (b) the practical, as the base for planning the national economy.

L.D. Stamp pioneered the land use studies in Britain. His monumental work (Stamp, 1951)[12] have encouraged and provided guidelines to geographers all over world. In 1930, Stamp established an independent research organisation called "Land Utilization Survey of Britain". The objective of the survey was to record the use of every parcel of land in Britain. It was decided to collect information on topographical sheets 6 inches to a mile which showed individual field boundaries. Such maps were available in size 18″ by 12″, quite handy for field work. Each sheet represented about 6 square miles of area. Appeals were then issued for volunteers. A large number of students from schools, colleges and universities came forward to take up the land use work. The enthusiasm ran so high that some volunteers even paid for their maps and postage.

Almost the whole of Britain was mapped before the outbreak of the War in 1939. The results were then checked and edited in a series of coloured land use maps on 1" to a mile and reports were completed by 1946. There were 92 reports, which covered almost 5800 pages. Later on a voluminous book *The Land of Britain : Its Use and Misuse* (Stamp, 1962)[13] incorporating the findings of the survey was published.

During the War of 1939-45 Great Britain was threatened with acute food shortage. The results of land use survey helped much in the Grow More Food campaign. The food production rose by 70 per cent during the War. Hence Stamp's efforts were rewarded and the usefulness of the land use survey was established beyond any shadow of doubt. A second land use survey of England and Wales, on a scale of 1 : 25000 has been recently undertaken under the direction of Miss A. Coleman and K.R.A. Maggs and numerous sheets and reports have been published (Coleman, 1961)[14]

In the post-war period the study of land utilization has been undertaken by many universities and research organizations. In some countries aerial photographs are used for preparing the land use inventories. The mapping technique has improved and new dimensions have been added in the study (Sharan, 1969).[15] Today one can find excellent land use maps of Cyprus, Italy, Poland, Hong Kong and parts of Canada, the U.S.A. and France.

Land utilization maps of Cyprus (1 : 2,500,00) have been prepared under the direction of R.R. Rawson and K.R. Sealy in the Department of Geography, London School of Economics in 1956. The maps were prepared from about 10,000 air photographs (1 : 10,000 to 1 : 13,000) taken in 1949 (Melamid, 1958).[16] The land use maps of Italy, are being published representing Calabria. It is in tune with the implementation of the decisions taken by the Italian National Research in the early 1950s to participate in the drawing of a world map of land utilization. As a result of recommendations made by the Commission on World Land Use Survey of the International Geographical Union, the Italian delegation to the 1952 International Geographical Congress presented a map, on the scale of 1 : 200,000 dealing with land utilization in Tuscany. The task of preparing the series was entrusted to the centre of studies in Economic Geography at the University of Naples and design scale and other details were worked out by a Committee composed of the leading Italian specialists in agricultural economics and economic geography (Kish, 1968).[17]

In Poland under the direction of J. Kostrowicki (1968), Department of Geography, Polish Academy of Sciences developed a new pattern of land utilization based on agricultural typology, agricultural regionalisation and planning or programming agricultural development (Kostrowicki, 1968).[18]

The methods of land utilization vary in different parts of the world. It comprises the J.L. Buck Survey of China, L.D. Stamp works in Britain and the surveys done in the United States either at regional or at state levels. The objective of these surveys and the methods followed are quite different from one another as they have been conducted under varying conditions prevailing in the area of study. It would be worthwhile to review briefly the purpose of these surveys.

The land utilization survey of Britain was a voluntary organization under direction of L.D. Stamp, whose aim was to survey the whole country, with a view to finding exactly for what purpose the land of the country was used. The results of the survey gave an assessment of the factors responsible for the complex pattern of land use in Britain. The main emphasis was laid on the classification of land use.

The survey done in the United States, on the other hand, are land capability surveys, as their purpose is to record, not only the use of the selected unit of the land at one time, but also to indicate the best suited use of the land. The purpose of Buck's survey was threefold: (a) to train the students in the methods of research in land utilization, (b) to make available knowledge of China's agricultre, and (c) to arouse interest among people of other countries in China's welfare, land utilization, food and population problems.

In China, data were collected on sampling basis from 22 provinces. Besides farm studies, other aspects related to agriculture such as food, standard of living of the people, marketing and price level were also considered. However, in the whole study no attempt was made to record the use of land on maps, which is an important aspect in land use study.

The entire study which runs to 470 pages contains 206 tables, 62 photographs and only 21 maps, which are mainly devoted to geographical background and China's agriculture.

History of land use study in India : Land use studies conducted by Indian geographers in various parts of the country received inspiration from L. Dudley Stamp,[19] who had attended the 25th session of the Indian Science Congress at Calcutta in 1938. "Such studies range from inventories of land use surveys to isolated topical or regional descriptive accounts of land use variations, both in space and time. Recently the studies are shifting towards the application of quantitative techniques in the analysis of various land use components".[20]

The land utilization survey of 24 Parganas[21] and Howrah districts[22] conducted by S.P. Chatterjee and the land utilization survey in Eastern Uttar Pradesh[23] conducted by M. Shafi have made a strong plea to carry out the land use survey combined with the survey of land capability. This will help in determining the best utilization of the land in relation to their intrinsic qualities. It was in the year 1940 that S.P. Chatterjee[24] tried to organise the land use survey in India. In his presidential address before the geography and geology section of the Indian Science Congress Association, 1940, he pointed out the necessity of understanding the land use survey. The Commission on World Land Use Survey pursued this matter and the Government of India established a National Committee for the purpose under the direction of S.P. Chatterjee. Prior to this, a scheme was drawn for preparing land use maps on scale 1 : 1,000,000 as a part of the National Atlas Organisation. By that time, no systematic land use survey had been initiated for entire country. Some geographers were engaged in pilot surveys individually. S.P. Chatterjee surveyed 800 villages of West Bengal and brought out eleven land use sheets on the scale of 4 inches to mile. Under the guidance of P. Dayal and A. Sharan, Department of Geography, Patna University, has surveyed three consecutive community development blocks, i.e. Biharsharif, Noorsari and Rahui in Nalanda district.[25] Besides this R.N.P. Sinha of Patna University intensively studied the land use of Canal Irrigated Area of Patna District.[26] In Bhagalpur University K.N. Das studied the land use of the Kosi Basin in North Bihar,[27] while S.R. Roy prepared his Ph.D. thesis of Land use and settlements in the districts of Bhojpur and Rohtas, Bihar.[28] The Government of Andhra Pradesh has conducted a land use survey of East Godavari district. Similar studies have been made by the Government of Karnataka for the whole State,[29] and for Thana Taluka by the State of Bombay. M. Shafi has brought out a scheme based on sampling techniques for land use survey of a vast country like India.[30] V.L.S. Prakasa Rao,[31] suggested land use classification on the lines of soil survey technique. He applied his idea to the land use analysis of the Godavari region. B.K. Roy has attempted an analysis of crop associations and changing the pattern of crops in the Ganga-Ghaghra Doab East.[32]

In Sagar University S.N. Mishra has studied land use in Khadar and ravines of the Lower-Middle Gomati Valley.[33] He has attempted land use planning for better adjustment of agriculture to the physical environment for optimum exploitation and conservation of natural resources. N.P. Ayyar[34] has given statistical approach to the study of crop-combination regions. He has suggested maximum distance method to group important crops into associations. H.S. Sharma[35] has tested the method of land capability classification based on Bennetts' method. According to Learmonth and Bhat[36] agricultural land use has been studied in detail by several disciplines but its sharp regional contrasts need spatial analysis which falls under the purview of economic geography. The Agricultural Atlas of India (1958),[37] the National Atlas (1957)[38] and the Census Atlases (1961)[39] of different states contain choropleth and dot maps relating to land use and crops which are useful in the analysis of crop distribution and concentration in general way and which have potentiality for more rigorous analysis.

In Ranchi University E. Ahmad[40] has analysed land use types in relation to physical elements. According to him slope of the land should be considered in preparing the development scheme of an Indian village. V.R. Singh[41] presented his Ph.D. Thesis on the Environs of Mirzapur Plateau in B.H.U. and calculated the land capability with potential productive unit and the standard nutrition unit. B. Mandal[42] has elaborated the Weaver's method in analysing crop combination regions with special reference to North Bihar. In analysing the crop combination B.N. Sinha[43] has taken the help of Weaver's and Nelson's techniques.

P.C. Vat of Jodhpur has studied land utilization and farming efficiency by applying the British method of land utilization. J.S. Yadav[44] has interpreted the broad regional variations in crop land use in Rajasthan. L.S. Bhat[45] has suggested the concept of planning from below, i.e., village, block and district level for land use planning. S.S. Sharma stressed the physical and human factors for determining land use patterns in Shahabad Tahsil of Mathura. V.B. Tripathi[46] of Kanpur and S.P. Garg[47] of Dehradun have applied Weaver and Doi's method to the study of crop combination ranking and changing patterns. B.N. Jha[48] studied the problems of utilization in the Kosi basin, which passes through the various physical and cultural activities including the emerging problems of land utilization in the region.

K.K.L. Das[49] studied about "Population and Land Resources in North Bihar Plain—West of the Kosi" in which he has shown the relationship between population and the intensity of the utilization of land in the study area.

S.L. Das[50] studied the "Patterns of Population and Land Use in the District of Bhagalpur, Bihar, 1971" in which he has shown the distribution, growth and future estimate of population, and agricultural land use in different *anchals* of the district. In this way he has investigated the pressure of population on land after considering various attributes of the environment.

The ultimate goal of land use study is to suggest the planning for better utilization of available land to the society. Land use planning is not recent in its origin but its practice is truly recent. A land use map gives a clear picture of land to the planner who determines its future use and suggests the maintenance of land potentiality.

Land use planning should have a humble beginning from the present position of the land use survey, so that the future use must satisfy the basic deeds of mankind e.g., food, shelter and cloth.

Sources of data : Materials have been collected from numerous sources and they include both published and unpublished maps and literature, census data, unpublished data from official sources, and personal field investigation.

1. 1″ to 4 miles (1 : 253,440) District maps, Published by State Survey Office, Gulzarbagh, Patna.
2. 1″ to 4 miles Survey of India (SOI, Dehradun).
3. 1″ to 1 mile Topographical Sheets of SOI.
4. 16″ to 1 mile (1: 3960) Cadastral Maps for the villages.
5. Various other maps contained in SOI Atlas, Rennell's Atlas, Dayal's Bihar in Maps, Learmonth & Bhats', An Atlas of Resource—Mysore State; Chatterjee's Bengal in Maps and Census Atlas of Bihar (S.D. Prasad), Office of the Census Operations, Patna.
6. *Data* : The data available in published form were the Census figures from 1951 to 1971 including village level data in District Census Handbooks of different districts. The land use data, agricultural statistics of crops, harvests and average yields at the district level were obtained from the Directorate of Statistics and Evaluation, Patna; various other data on sample villages were obtained from the Karamcharies, V.L.Ws.; and Mukhias by the author during field work, due care in assembling, sorting, cross-checking and computing of these has been taken before use.
7. *Field Work* : The author was acquainted with the region on account of his earlier work on Bihar Plain. In course of the present work, he visited the region several times between 1971 and 1977 for the collection of relevant data, personal observation, etc. The sample villages were visited during different harvests with this end in view, most of his vacations viz. Summer, Puja, Winter and Holi were devoted to field work. Besides field work in the area of study, a good deal of information on important aspects of the problem were collected.

Hypotheses Formulation : Before dealing with the hypothesis formulation, it is necessary to know about the general framework for carrying out any research projects in the field of agricultural geography. In order to their presentation the steps are as follows:[51]

1. Selecting the problem of research,
2. Stating the problem,
3. Thinking through the purposes,
4. Formulating the hypotheses,
5. Determining the evidence needed,
6. Assembling the evidence,
7. Processing and analysing the evidence, and
8. Presenting the results.

All these are the different steps of research and in this sense one should not go ahead on to the second step, without clearly understanding the first. Thus 'step' is a series of actions in which the previous step is an undertaking. But a research activity might not complete all of the steps, it might even stop at the point of the formulations of hypotheses.

The hypothesis shows the relationship between two or more variables. Thus a statistical hypothesis is a formidable answer to a question about relationships. Statistical hypotheses are usually classified into null hypotheses and alternative hypotheses and test of such a hypothesis consisting of testing the null hypothesis against the alternatives or research hypothesis. The hypothesis $\bar{x} \neq \bar{x}_1$ is tested against the alternative $\bar{x} = \bar{x}_1$. Where $\bar{x}$ and x_1 are the means of two observed variables, whereas $\neq$ is a sign of not equal to. Hypotheses originate from the observations of events. They provide a bridge between the theory and observed real world situation. This is because researchers generally ask questions about the relationship between variables in the light of the existing theory. After the formulation of the hypothesis, it is advanced for testing and drawing inference.

The testing of hypothesis may proceed by drawing deductive or inductive logic of inferences. In deductive research events are explained by citing conclusions which are based upon theoretical principles, whereas in inductive research relationship between two variables is illustrated on the basis of data gathered through experience. This is because we have few principles and laws in geography and therefore, a geographical research is mostly inductive. Both theoretically and by common knowledge it could be suggested that population in a district should have close relationship with the net sown area in the district (1971). The problem can be restated by the hypothesis that the increase in population (dependent variable) is dependent on the increase in net sown area (independent variable). This simple relationship is illustrated in a scatter diagram (Fig. 1.1).

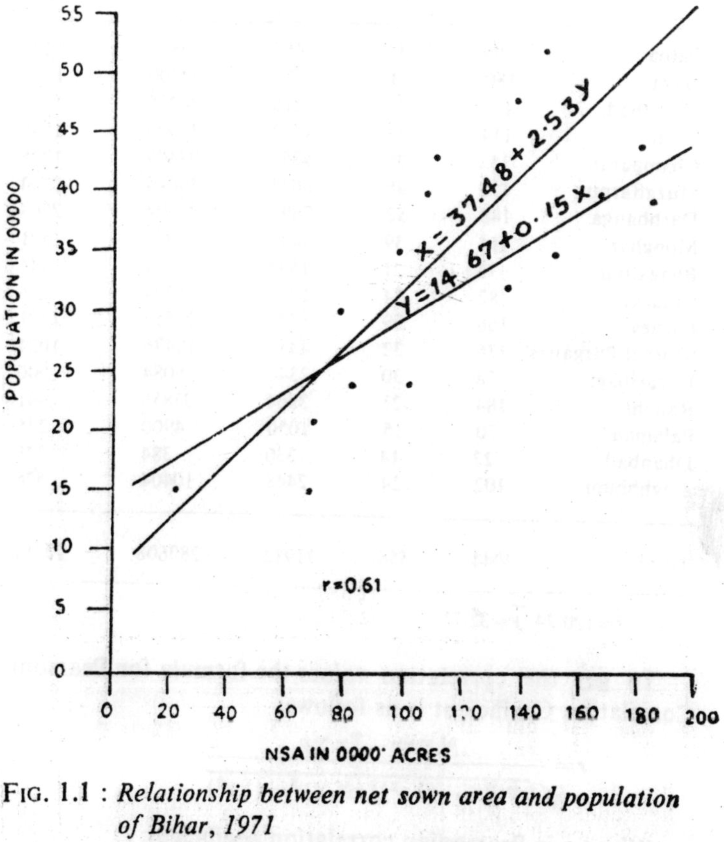

FIG. 1.1 : *Relationship between net sown area and population of Bihar, 1971*

In the scatter diagram the dependent variable has been shown on the *y* axis (vertical) and the *y* independent on the *x* axis (horizontal). To know the distribution and magnitude of one variable with the other comparisons may be made so that their correspondence can be estimated.

In order to test the hypothesis an example has been given below:

TABLE 1.1 : **Relationship between NSA and population in Bihar (1971)**

Districts	NSA in 0000' Acre	Population in 00000'	Xy	x^2	y^2
Patna	99	35	3465	9801	1225
Gaya	180	44	7920	32400	1936
Shahabad	185	39	7215	34225	1521
Saran	114	43	4902	12996	1849
Champaran	153	35	5355	23409	1225
Muzaffarpur	138	48	6624	19044	2304
Darbhanga	148	52	7696	21904	2704
Monghyr	114	39	4446	12996	1521
Bhagalpur	73	21	1533	5329	441
Saharsa	82	24	1968	6724	576
Purnea	166	39	6474	27556	1521
Santhal Parganas	136	32	4352	18496	1024
Hazaribagh	78	30	2340	6084	900
Ranchi	184	21	3864	33856	441
Palamau	70	15	1050	4900	225
Dhanbad	22	14	330	384	225
Singhbhum	102	24	2448	10404	576
Total	2044	556	71982	280608	20214

$\bar{x}=120.24, \bar{y}=32.71$

To get the correlation values the formula for Pearsonian Correlation Coefficient is as follows:

$$r=\frac{N.\Sigma xy-\Sigma x.\Sigma y}{\sqrt{(N.\Sigma x^2-(\Sigma x^2))\,(N.\Sigma y^2-(\Sigma y^2)}}$$

Where r is Pearsonian correlation coefficient,

Σx is sum of all values of independent variable,
Σx^2 is sum of square of all independent variable,
Σy is sum of all values of dependent variable,
Σy^2 is sum of square of all dependent variable,
Σxy is sum of product of all x and y values, and
N is number of units.

With the help of above table the product moment correlation coefficient calculation is as follows:

$$r=\frac{17\times 71932-2044\times 556}{\sqrt{(17\times 280608-(2044)^2\,(17\times 20214-(556)^2}}$$

$$=\frac{1223694-1136464}{\sqrt{(4770336-4177936)\,(343638-309136)}}$$

$$=\frac{87230}{142912}=0.61\,(r)$$

Calculating from the above, we get the value of $r=0.61$. This shows that the relationship is very close. The result has been tested by students '*t*' test by the formula:

$$t=\sqrt{\frac{r^{\cdot}(n-2)}{1-r_2}}=\sqrt{\frac{.3721\,(17-2)}{1-.3721}}$$

$$=\sqrt{\frac{5.5815}{0.6279}}=\sqrt{8.88}=2.98$$

Ho: According to Null Hypothesis (*Ho*):

There is no correlation between net sown area and population in different districts of Bihar (1971).

Hi: According to alternative hypothesis (*Hi*):

There is correlation between net sown area and population in different districts of Bihar (1971).

Table value of '*t*' at 15 degree of freedom (*df*) for 50 per cent and 1 per cent level of significance are 1.753 and 2.602, whereas calculated value of $tc=2.98$. Since the calculated value of '*t*' is significant at 5 per cent and 1 per cent level of significance at *df* 15, therefore *Ho* (Null Hypothesis) is rejected in favour of *Hi* (Research Hypothesis).

The regression of *y* upon *x* and *x* upon *y* have been calculated by the method of least squares according to the formula:

$y=a+b\,x$ and $x=a+by$.

where *y* is the dependent variable and *x* is the independent variable. The values of *b* and *a* are given by the following formula:

$$b=\frac{N.\Sigma xy-\Sigma x.\Sigma y}{N.\Sigma x^2-(\Sigma x)^2}\text{ and }\frac{N.\Sigma xy-\Sigma x.\Sigma y}{N.\Sigma y^2-(\Sigma y)^2}$$

for *y* and *x* respectively.

Similarly $a=y-x.\bar{b}$ for *Y*, and $a=\bar{X}-\bar{Y}.b$ for *x*.

where $\bar{x}$ and y parameters are the same as mentioned earlier.

In the present case, the x and y values have been estimated as:

$X=37.41+2.53\ Y$ and

$Y=14.67+0.15\ X$

The accuracy of this estimate of y and x can be tested by calculating the standard error of estimate. It is a measure of the error to be expected in estimating the value of y from a given value of x, and the value of x from a given value of y by means of the computed value of y and x respectively. The formula is as follows:

$$\text{Standard Error (Sey)}=\frac{1-r^2}{\sqrt{n}}\ \sqrt{\frac{1-.3721}{17}}$$

$$=\frac{0.6279}{4.12}=0.15$$

The standard error of estimate gives the probability of error of two-thirds of all cases on either side of the regression line. Thus the final equation takes the form:

$Y=14.67+0.15\ X\pm0.15$

$X=37.48+2.53\ y\pm0.15$

Fortunately, the standard error of estimate is very low.

Need of hypotheses testing in research: To improve the methods of research in geography, hypotheses may be stated and tested. In geography, the techniques of mapping, photo-interpretation, remote sensing and field work enriched the analysis of spatial distributions of crops. All these enable the investigators to observe and record accurately so that hypotheses can be tested against real world situations.

Recently several new techniques have been adopted for hypothesis testing in geographic literature. In 1960's regression techniques were quite common but today factor analysis is very popular among geographers. Very recently, multidimensional scaling has been introduced to investigate into the problems related to the choice between alternative locations. These techniques have been borrowed from social sciences to geography to explore relationships implicit in spatial organization.

The need of hypothesis testing is to improve the theoretical constructs of the geographic discipline. Thus in order to know the exact relationship between theory and occurrences, hypotheses must be developed by those who have sufficient knowledge of the past. Generalisations published in previous literature must be tested under different cultural circumstances.

Lastly it can be said that in hypothesis formulation answerable questions are examined with the help of field observation, mapping and review of past literature. All these help in refinement of propositions and maximization of research efforts. Hypothesis formulation also represents the initial step in a well established approach of research.

REFERENCES

1. T.W. Freeman, *Geography and Planning*, Hutchinson University Library, London, 1968, p. 74.
2. J.N. Jackson, *Survey for Town and Country Planning*, Hutchinson University Library, London, 1963, p. 109.
3. Quoted by M.B. Nanavati and J.J. Anjaria, *The Indian Rural Problem*, Vora & Co., Publishers Ltd., Bombay, 1951,
4. M.M. Kelso, Scope, content and orientation of rural land economics research today, Land Economics Research (Edts. Ackerman et al.), *Farm Foundation, Resource for Future, INC.*, 1962, p. 20.
5. M.B. Nanavati, (Foreword), *Readings in Land Utilization*, The Indian Society of Agricultural Economics, Bombay, 1957, p. 2.
6. R.G.S. Stapledon, *The Land: Now and Tomorrow*, 1940, p. 32.
7. M.M. Kelso, op. cit., 1962, p. 21.
8. M. Shafi, Can India support five times her population, *Science Today*, May, 1969, p. 24.
9 Soil series may be defined as "a Group of soils having in general, the same character of profile—that is, the same range in colour, structure, and general sequence of soil and subsoil horizons, the same Type of Relief and Drainage and a common or similar Origin or Mode of Formation". Show, C.F., A Glossary of Soil Terms, *American Soil Survey Bulletin* 9, 1927, pp. 28-58.
10. *Readings in Land Utilization*, The Indian Society of Agricultural Economics, Bombay, 1957, p. 43.
11. A. Sharan, Land use and planning, *Essays in Geography* (Edt. K.N. Das) Bhagalpur, 1969, p. 32.
12. L.D. Stamp, *Land for Tomorrow:The Under-Developed World*, Bloomington, 1951.

13. L.D. Stamp, *The Land of Britain : Its Use and Misuse*, London, 1962.
14. Alice Coleman, The second land use: Progress and Prospect, *Geographical Journal*, Vol. CXXVII, 1961, pp. 168-73.
15. A Sharan, Land use and Planning, *Essays in Geography* (Edt. K.N. Das) Bhagalpur, 1969, pp. 30-33.
16. Alexander, Melamid, "Land use in Cyprus" *The Geographical Review*, 1958, Vol. XLVIII, No. 1, pp. 112-113.
17. George Kish, A new land use map of Italy, *The Geographical Review*, Vol. XLVIII, No. 1, 1968, p. 270.
18. J. Kostcowicki, Agricultural typology, agricultural regionalization and agricultural development, 21st I.G.C. Abstract Papers, Calcutta, 1968.
19. FAO; agricultural commodities, projection for 1975 and 1958, (1967).
20. M. Shafi, *Land Use Studies: A Trend Report*, A Survey of Research in Geography, Popular Prakashan, Bombay, 1972, p. 19.
21. S.P. Chatterjee, Land utilization in the District of 24 Parganas, Bengal, 1945, *B.C. Law*, Vol., Pt. 2, Calcutta, 1945.
22. S.P. Chatterjee, Land utilization survey of Howrah district, *Geographical Review of India*, 14 (3) 1952.
23. M. Shafi, *Land Utilization in Eastern Uttar Pradesh*, Aligarh, 1960.
24. S.P. Chatterjee, The place of geography in national planning, *Calcutta Geographical Review*, 3, 1941.
25. P. Dayal, and A. Sharan, Land use patterns in the Biharsharif Area, Patna District, Bihar, *Proceedings of Symposium on Land Use in Developing Countries*, 21st IGC, Aligarh, 1972, pp. 146-158.
26. R.N.P. Sinha, Present land use pattern in canal irrigated area of Patna District, Bihar, *Geographical Outlook*, 4, 1965.
27. K.N. Das, Population and Land Use Changes in the Kosi Region Bihar, Unpublished Ph.D. Thesis, Bhagalpur Universlty, 1969.
28. S.R. Roy, Land use and settlements in the Districts of Bhojpur and Rohtas, Bihar, Unpublished Ph.D. Thesis, Bhagalpur University 1976.
29. "Land Utilization in Karnataka", Evolution Unit, Planning Department, Bangalore, 1975.
30. M. Shafi, Techniques of rural land use planning with reference to India, *Geographer*, 13, 1966.
31. V.L.S.P. Rao, Land use survey in India: Its scope and some problems, *Proceedings of the International Geographical Seminar*, Aligarh, 1956.
32. B.K. Roy, Crop Associations and changing pattern of crops in Ganga-Ghaghra Doab East, *National Geographical Journal of India*, December, Vol. VIII, No. 4, 1967, pp. 194-207.
33. S.N. Mishra, Land use in the Khadar and ravine tract of the lower

middle Gomati Valley, *National Geographical Journal of India*, 10 (3 and 4), 1964.

34. N.P. Ayyar, Crop regions of Madhya Pradesh: A study in methodology, *Geographical Review of India*, Vol. XXXI, No. 1, March, 1969, pp. 4-5.
35. H.S. Sharma, Land capability classification of the lower Chambal Valley, *Proceedings of Symposium on Land Use in Developing Countries*, 21st IGC, Aligarh, 1972.
36. L.S. Bhat, and A.T.A. Learmonth, Recent contribution to the economic geography of India: Some current preoccupations, *Economic Geography*, Vol. 44, July, No. 3, 1968, pp. 189-209.
37. Government of India, Agricultural Atlas of India, New Delhi, 1958.
38. S.P. Chatterjee (ed.), *National Atlas of India* (preliminary Hindi Edition), 1957.
39. (a) Registrar General of India: *Census of India*, 1961, State Atlases of Andhra Predesh, Bihar, Gujarat, Uttar Pradesh, etc.
    (b) Registrar General of India: *Census of India*—1961, Vol. 1 part 1A (i) Levels of Regional Development in India.
40. E. Ahmad, *Geographical Essays on India*, Patna, 1954, pp. 15-16.
41. V.R. Singh, *Land Use Patterns in Mirzapur and Environs*, Ph.D. Thesis, Banaras Hindu University, 1970.
42. B. Mandal, Crop combination regions of North Bihar, *The National Geographical Journal of India*, Vol. XV, Part II, June, 1969, pp. 125-37.
43. B.N. Sinha, Crop combination technique: A search for an Ideal Tool, *Deccan Geographer*, 6 (2), 1968.
44. J.P.S. Yadav, Crop land use patterns in Rajasthan, *Conference of Geographers*, Bhilwara, 1965.
45. (a) A.T.A. Learmonth & L.S. Bhat, (Edts.) *Mysore State*,—Vol. 1, An Atlas of Resources, Asia Publishing House, London, 1960.
    (b) A.T.A. Learmonth, & L.S. Bhat, (Edts.): *Mysore State*, Vol. II, An Atlas of Resources, A Regional Synthesis, 1962.
46. V.B. Tripathi & U. Agarwal, Changing patterns of crop land use in the Lower Ganga Yamuna Doab, *Geographer*, 15, 1968.
47. S.P. Garg, Land utilisation in Saharanpur District, Unpublished Ph.D. Thesis, Agra University, 1968.
48. B.N. Jha, *Problems of Land Utilization: A Case Study of Kosi Region*, Classical Publications, New Delhi, 1980.
49. K.K.L. Das, Population and Land Resources in North Bihar Plain-West of the Kosi, Unpublished Ph.D. Thesis, Bhagalpur University, 1979.

50. S.L. Das. Patterns of Population and Land Use in the District of Bhagalpur, Unpublished Ph.D. Thesis, Bhagalpur University, 1979.
51. M. Harris, "Steps in carrying out formal research project, *Land Economics Research*, Ackrmoan, J. et al. (Ed.) Washington, 1962, p. 231.

2

Factors of Land Use

LAND USE is conditioned by the association of two sets of factors (Fig. 2.1)—first, physical factors such as geology, relief features, climate, soil and vegetation which limits the use capabilities of land and secondly, cultural factors which include both economic and institutional factors. Cultural factors represent the length of occupance of the area, demographic and socio-economic conditions, institutional framework and the technological levels of the people which determine the extent to which the land can be utilized.[1]

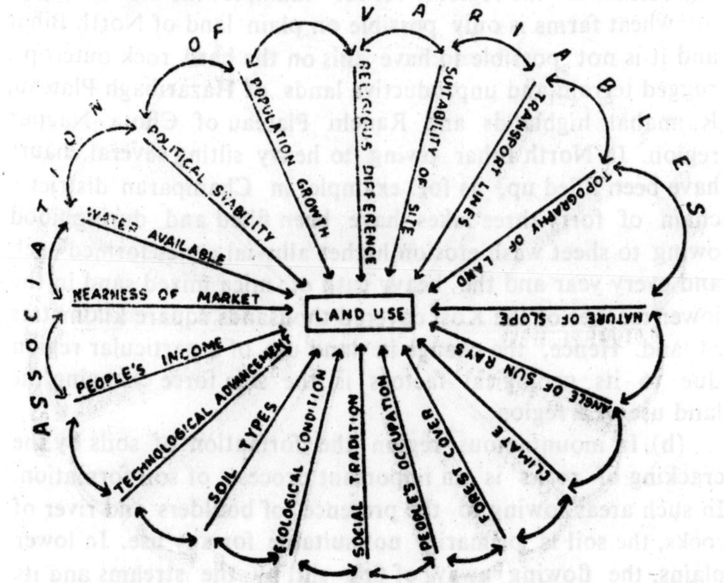

FIG. 2.1 : *Ecological Balance of Land Use*

Thus, in general, there are three factors of land use, e.g. (1) Physical factors which include geology, relief surface, drainage underground water, climate, soil and vegetation. (2) Economic or Locational Factors, (3) Institutional Factors.

PHYSICAL FACTORS

Geology : Geological structures have both indirect and direct influences on land use. The indirect influences may be categorised into:

(a) ground having presence of absence of any geological occurrences,
(b) formation of soil from the parent source region,
(c) influence on permanent water table, and
(d) surface water supply and underground drainage.

(a) Land utilization of any region is the direct product of its geological structure and man is the main source behind moulding land use, according to their need and the geological ingredients of the region. As for example the availability of big wheat farms is only possible on plain land of North Bihar and it is not possible to have tnis on the hard rock outcrops, rugged terrain and unproductive lands of Hazaribagh Plateau, Rajmahal highlands and Ranchi Plateau of Chota Nagpur region. In North Bihar owing to heavy silting several chaurs have been filled up, as for example, in Champaran district a chain of forty-three lakes have been filled and during flood owing to sheet wash erosion higher alluvial cones formed each and every year and the heavy rush of mica mixed sand in the lower reaches of the Kosi covered thousands square kilometres of land. Hence, the change in land use of a particular region due to its geological factors is the sole force changing the land use of a region.

(b) In mountainous region the formation of soils by the cracking of rocks is an important process of soil formation. In such areas owing to the presence of boulders and river of rocks, the soil is primarily not suitable for any use. In lower plains, the flowing away of fine silt, by the streams and its deposition in the form of alluvial soil is of great importance for a judicious land use scheme. In case the parent rock is calcareous then in the lower plains or valleys calcareous soil will form and in case of laterite rocks on the higher land, in lower areas red soil will form. Originally, the areas of sand dunes are not suitable for crop raising but with the help of irrigation good crops may be raised e.g. in the Punjab, Rajasthan and Haryana areas of India.

(c) The texture of drainage and availability of underground water are the products of their geological conditions. Underground water is generally available in three types of rock structures, e.g. in acquifer, sandy layer and in calcareous rocks, because these are sufficiently porous to hold water. These rocks also allow the passage of water through them, if there is lack of impervious layer of rocks in between the two continued beds. There are three sources of obtaining underground water (a) surface well, (b) deep tube-wells and (c) springs.

Surface well : In the surface well or shallow wells the supply is from the sandy layer or acquifers near the surface in the bed of gravel.

Deep tubewells : In deep wells the water supply is mainly from acquifer or sandy layer at great depths. In case the rocks are arranged in the form of a basin the supply water may be huge in amount. After shipage through impervious rocks fountain results giving rise to artesian wells. In Australian artesian basin and London basin of England such wells are numerously found but in London basin, the artesian wells are now not giving water, naturally, but the pumping process is necessary. So far as the water giving capacity is concerned the difference is clearly visible in the Chalk country of England. In case of fissured Chalk, water comes out freely but in case of hard rock the ejection of water is low.

Springs : When underground water comes on the surface by means of shipage through impervious rock structure it is known as springs. To mould the land use of any region this source of water has great value in the respective areas. There were 1,000 deep wells in the Santa Clara Valley of California in 1910 but by the end of 1930 all the wells dried up and the ground gradually and slowly started sinking at the rate of 5 feet in 20 years time. Hence, this example is to be kept in mind, especially in India where the sinking of tube-wells is extensively going on in the northern plains without assessing the capacity of underground reservoirs. In this process if more water is drawn than its annual natural supply by rainfall, etc. there may rise a danger of the drying up of shallow wells and permanent depletion of underground water. It should be used judiciously, because water is not available abundantly on the upper surface. Assured water supply is essential for market gardening, industrial location, domestic pursuits, raising of crops in the dry season and the availability of water is largely dependent upon the local underground water reservoir or perennial streams. In desert region the oasis is a good example of water concentration and the utilization of land is only available on such sites of underground water resource.

(d) There are two types of drainage: (i) surface and (ii) sub-surface. In areas of heavy rainfall quick run off over steep slope leads to heavy soil erosion and this makes the land unfit for any use. In areas of higher land slope gradient, afforestation work and planting of grasses may save the land to some extent but the major portion becomes useless. Improvement of drainage on higher slope may increase the danger of flood in the lower reaches but this requires watershed planning. On account of this, in India it is necessary to boost afforestation work especially in the Terai areas of Nepal to check the menace of flood and to improve utilization of land in the northern plains of India. It also reduces the problems of flood, checks the higher amount of silting and regulates the supply of water throughout the year. This also requires the presence of straight channels having no meandering courses in the lower reaches near the Ganga for the huge discharge of Himalayan rivers during rainy season. Owing to this fact, deterioration in the standard of land use is observed in the flood plains of riverine belts.

In areas of heavy soil like clay, the digging of canals for quick runoff water is necessary to keep up the standard of land use. The character and texture of drainage determines the land use in any region and in turn the condition of drainage is dependent upon the structure or rocks.

GEOLOGICAL STRUCTURES HAVING DIRECT INFLUENCE ON LAND USE

The mining areas of iron ore, coal, limestone, etc. are the zones of low land use as the piles of coal, network of transportation lines and highly rugged terrain restrict surface utilization of land up to its mark. Besides this in mining areas the problem of subsidence of ground, large scale disruption of settlements, wastage of land in odd works, heavy rushing of underground water are some of the disastrous problems which affect the utilization of land in any mining areas. In Bihar the disaster of Chasnala mine is one of the memorable tragedy of this sort in the mining history of the world. Similar cases of subsidence may occur in salt mining areas. In slate quarrying areas, owing mixture of slate pieces in the soil, this restricts even the germination of vegetation for a longer period of time.

The filling of the holes of mineral extraction is a gignatic task. This may be a problem of a smaller area but in the interests of the nation, the replenishment of such land is highly essential for the future judicious use. In this direction the beginning of mechanised mining and the filling of sand is necessary to check large scale subsidence and malutilization of land, especially in areas of agricultural interest.

SOIL AND LAND USE

Soil is the basic determinant of land use in any region. In almost all plain areas of the world, especially in river basins alluvial soil is predominantly found, which is conducive to holding water and producing good crops. In comparison with this, in plateau and hilly areas coarse sandy soils with a mixture of stone pieces or red soils are found which are not suitable for raising bumper crops. In the areas of waterlogging generally saline soils are found which are unfit for raising food crops abundantly. In desert areas, mostly sand dunes are found in which only the sites of oasis are amenable to better utilization of the land and the rest are useless for lack of surface water or irrigation.

After knowing the effects of soil types on land use the characteristic features of soil such as the determination of PH value is a very nice determinant of both soil types and land use. Such as a PH value of 4 in any soil means the presence of acid in the soil, PH value of 7 means neutral soil. In most of the agricultural areas, the PH value in soils varies from 5 to 5.8. What type of fertilizer should be used in a soil is largely dependent upon the PH value of the soil. Besides this, for the healthy growth of plant life 17 chemical elements are necessary which is generally known by chemical analysis and the character of soil may be changed by the use of fertilizer and tillage practices. In view of this U.S. Department of Agriculture has grouped the soil of the U.S.A. into eight categories according to its capability and suitability of land use. In this classification, categories first to fourth are suited for crops, pastures, woodland and wild life and categories 5th to 8th suited for pasture, woodland and wild life, but are not suitable for cultivation.

Land tillage varies according to the nature of the soil and the purpose as follows:

(a) to change the structure of soil,
(b) to kill weeds, and
(c) to manage crop residues.

Geological structure of a land determines the method of tillage. In case of shallow soils on the upper layer with hard basement deep furrowing is not suitable and, therefore, one has simply to use hoe, spades, etc. Deep layer of alluvial soil alone permits deep furrowing by heavy implements.

The rotation of crop is.also necessary for the continuous maintenance of soil fertility. Thus, it is advisable to use specific type of crops on specific type of land, because in this way, land can produce good crops and add nitrogen and humous, etc. to the soil according to the nature of crops.

Soil management needs the categorisation of land in accordance with capability. The practice of soil conservation should be adopted after assessing the geological structure of the land, physical and chemical properties of soil and the need to conserve soil which are immune to erosion. In this way the steep slopes may be left for afforestation work, slopes on the hillsides for raising grass to breed cattle and in the plains the agricultural operations should be permitted.

RELIEF FEATURES

The intensity of land use and the distribution of population show that 75 per cent population of the world live in plain land up to 1000 feet from mean sea level. This is due to the availability of sufficient amount of water for the performance of agricultural operation in comparison with the desert environment. But all plains are not equally habitable such as the Arctic, frost-frozen ground, tropical dense forests, deltas and coastal plains have contained excessive moisture in the air. In such areas the problems of surface and underground drainage area always there to determine the extent of land use. In such an environment the arable farming is difficult and the ripening of crops becomes uncertain. In these areas, swampy ground, marshes, grass and thick forests are found. In order to make the land habitable, people have cleared forests, circumvented hills, constructed river-embankments, reclaimed swamps and marshy areas for the adjustment of resourceful land use. Further, some examples have been given here to determine the possibility of limitations in land use.

Coastal plain : The poor drainage and unfertile soil are the two hinderances of proper land use in the coastal lowlands. Here the drainage is related to slope and unfertile soil to the sandy waste. In this way on Malabar, Coromandel and Konkan coasts wilderness still prevails along with hunting and commercial development of ports. In all around the port site, intensive land use is found.

Deltas : In deltaic plains, the characteristic feature is the poorly drained land in between levees where the soil is fine and unconsolidated. The rivers are generally divided into numerous sluggish channels with abnormally wider course and the unfertile sand deposits abounding in sand bars and levees. Here, the agricultural plots are generally found in a linear fashion along narrow embankment away from the swampy ground. Deltaic plains are mostly sparsely settled but high population density is also found as in the Ganga delta and the Nile delta. Northern side of Sunderban delta is an example of reclaimed deltaic marshy ground for better land use.

Flood plains : Flood plains are the products of rivers coming from the higher ground and such rivers deposit vast amount of sand and silt especially during flood. The meandering streams and oxbow lakes are found, as these are the signs of senile stage for rivers to meet with the deltaic plains. Owing to lesser variation in slope, the problem of surface drainage is low and owing to periodic renewal of alluvial soil, excellent agricultural regions are formed. On such plains bumper crops are raised, roads built, embankment laid down on the side of rivers and the locations of towns and rural settlements are found above normal flood level. North Bihar is a good example of flood plains in India, where the general land use is determined by the availability of water, meandering streams, river embankments and permanent water-bodies e.g. Kabartal and Barailatal in Vaishali and Begusarai districts of North Bihar respectively. Oxbow lakes are also found in Vaishali district along the Burhi-Gandak river. Similarly, marshy grounds are numerously found in the Kosi flood plain of Saharsa district.

Piedmont plains : The materials deposited on such plains range from coarse granular to fine sands and gently sloping plains are found away from the piedmont area. In moist region, this is best suited for agricultural use but for rain-shadow zones, irrigation is much needed.

Glaciated plains : The source of the soil is found in mountainous belt from where the coarse material drifted downwards and deposited in the form of glaciated morains, moors, terminal morains along with the deposit of smooth silt and sand. Generally, forest cover instead of good agricultural land use is found then Glaciated plains of North America and Kashmir valley in India, good agricultural land use is found there but morainic area is preferred for livestock raising and dairying.

PLATEAU AND LAND USE

Plateaus are higher horizontal land mass, evolved due to tilt or uplift and are composed of hard rocks. In youth stage this may be a featureless flat topped country but in maturity plateaus may become badly dissected. Two plateaus are often separated by narrow steep sided valleys. It is not suitable for farming and in most cases plateaus are erosional surfaces. Only on rolling plateau deep layers of soils are found and it is for this fact that it becomes suitable for agricultural purposes. Plateaus are of three types :

(a) mid-Humid Latitude Plateaus,
(b) mid Latitude Dry Plateaus, and
(c) low Latitude Plateaus.

Owing to heavy rains humid plateaus show dissected topographic features such as Applachian, Ozark, Ouchita, Chota Nagpur, Shillong, etc. Such plateaus are mostly covered with forests or plantation crops where the production of food crops is not possible.

In mid latitude dry plateaus, higher percentage of grass land favours widespread animal rearing. Granchaco of Argentina and Australian Plateaus are the homes of sheep ranchers. Agricultural land use in such areas is possible only after irrigation is supported by glacial streams. Hay, wheat and sugarbeat are the important crops grown. The steep escarpments on the marginal belts of the plateaus restrict the construction of roads and railways.

Low latitude plateaus are mostly found in the tropical areas. In these areas subsistence agricultural practice is found. In the areas of European colonization e.g. in south-east Asia Plantation agriculture has also developed. Scattered patches of cultivable land are also found and that is controlled by the gradient of slope and depth of soil layers.

MOUNTAIN AND LAND USE

In mountainous country, hills and valleys are predominant factors in determining the intensity of land use according to slope. Thus agricultural practice on rugged terrain is totally impossible while terraced cultivation on small patches of land is the rule. Tracking of bulky animal is not possible on mountain slope, while donkeys, goats and sheep have adopted themselves to the slopes easily, on account of their light body. The stony-waste soil, rough terrain, dense forest cover and land restrict agricultural operation in the hilly region while in the tropics, sufficient amount of rainfall favours the cultivation of land on mountain slope, but the gradient of slope is the most determining factors of land use in the hilly region. The location of hill stations, points of scenic beauty, construction of road on the ridge, tunnels for railway siding and road are the important features of land use in mountainous region.

CLIMATE AND LAND USE

The climate determines the utilization of land. Under the impulse of climatic conditions the early distributions of dense population and the maximum use of land "seems to be confined approximately to a zone bounded by the Tropic of Cancer and the fortieth parallel of latitude. The climate is warm enough so that many plants can quickly complete their cycle of growth and take advantage of the interval between periods of seasonal rainfall".[2] Thus, under favourable climatic conditions, especially in India as many as four cycles of the growth of plants, viz., bhadai, aghani, rabi and garma corresponding to the rainy, winter, spring and hot weather may easily be discerned. In this way, it can be said that climate is the dominant factor of determining agricultural land use.

LOCATIONAL FACTORS

Land use is the product of varying factors in which locational factor has its own importance. Lands of economic importance are generally distributed around the centres of economic activity. Generally, costs are involved in moving the land products to the market through transportational lines and bringing capital and labour to the land. All these involve locational problem in land use study.

Man is the producer of goods on a particular piece of land and it depends upon him to select whether his earnings will be large, assured, and better working conditions available. Man is also consumers of such products and it depends upon him to select settlement site, where living is secure, cheap and easily agreeable. It is also found that a man living in any climate from generations to generations never wants to shift his settlements, this is due to his habitual living and utilization of land. All these enterprises and locational decisions are generally determined by the concepts of economic specialization and comparative advantge.

ECONCMIC SPECIALIZATION AND COMPARATIVE ADVANTAGE

Economic and functional specialization are common phenomena in the modern world. Individuals starting a business for themselves often perform a wide variety of tasks. Once their business expands and prospers, new employees are usually hired. This way, a division of labour takes place and each employee tends to do that type of work which he can do best in comparison with his fellow employers. Many of the principles that apply to individual specialization also apply to area specialization in production.

PRINCIPLE OF COMPARATIVE ADVANTAGE

The crops generally produced in local area have predominance of higher income in comparison with those in other areas. This is known as the principle of comparative advantage and owing to this fact, some areas specialize in the production of some crops, and look toward other area for so many products they use.

This principle can be best understood by taking the examples of two areas and two products respectively. Here we may assume that area C & D each produce maize and paddy crops according to their need and each area receives the net output per unit of input indicated below:

Crops	*Area C*	*Area D*
Maize	50	50
Paddy	55	55

In these cases not even a single area possesses production advantages, because they have no specialization in these crops.

In case of change in production situation in area *D* as indicated below, it is profitable for each area to specialize. Area *C* finds it profitable in producing paddy, while area *D* should concentrate on producing maize crops.

Crops	*Area C*	*Area D*
Maize	50	55
Paddy	60	50

In reality, there is occasionally some advantage for more than one use in any area while most areas fail to enjoy any absolute advantage for any use.

SCOPE OF COMPARATIVE ADVANTAGE

The comparative advantage is frequently sought in terms of natural advantages, e.g., favourable climate, soil and topography. The approach of comparative advantage is much more affected by the human judgment and policies. This may also be measured in terms of the economic ability of an area to compete with other areas in the production of given products within the control of environmental location, transportation cost and favourable institutional arrangements.

Natural advantages: The availability of minerals in sufficient quantity favours the use of land for mining. Salubrious climatic conditions in a region favours recreational developments. A port with productive hinterland favours urban development, e.g. Cochin, Kandla, Machhlipattam and others.

Favourable production combination : Natural advantage is known as comparative advantage in relation to the factors which enter into production. Shortage of skilled labour, capital, credit and marketing facilities may easily outweigh natural advantage of an area.

Transportation advantage : Owing to locational advantage in relation to transport lines, some areas have good linkage and in transporting goods from their area of produce to the market, generally, local producers get more benefit in moving the goods in lesser time and fresher condition than the outside competitors with their savings in the transportation cost. In supplying milk and vegetables to the market, this situation generally arises.

Institutional advantages : Institutional factors affect land use indirectly, such as tariff barriers, trade restrictions of field produce etc. as these hamper outside competition. This enhanced the production advantages enjoyed by the local producers such as protective tariff often gives definite market advantages to agricultural and industrial enterprises.

Principle of first choice: The supply of any products from an area shows the demand of land for a particular use and this is said to be the first choice of the area. This principle is applied for comparative advantages and the best use of land. Mining may be the first choice in areas of proved iron-ore deposits. Mountain passes may have first choice of railway line or roads and others. This situation arises due to higher demand and lesser supply of specialized use of land. Thus, from an individual point of view, a waterfall may be suitable for power site but the society may like to use it as the place of scenic attraction.

Spatial relationship of economic factors affecting land use: Land utilization of any region reflects locational differences with respect to market or point of consumption. This is particularly true when one deals with the land areas of the like productive capacity located at different distances from the market. Transportation cost is the key factor in these cases. The cost generally increase with the distance from the settlement site, hence the field located near the market has comparative advantage over the land located away from the market. This means that areas close to the market receive a higher advantage of price for the produce and thus yield more economic rent for their produce and have higher resource value than the areas located away from the market. This aspect of location factor is close to the idea of Von Thunen's concept.

In locational decision for a particular type of land use, these factors are constantly at work and producers often find that they must gear their operating decisions to national and international conditions. Instead of worrying about the local market demand for their products, businessmen often find their market in distant cities and towns. They never calculate their prospect of profit in terms of local competition but their market prices are influenced by the industrial employment, public price policies and changes in international supply and demand core.

INSTITUTIONAL FACTORS

The collective or social action of various groups that influences and controls the individual behaviour may be described as institutional factor. According to Griffith Taylor it is a purposeful cooperative behaviour.

It is often desirable to classify different types of institutional norms available in our society into primary and secondary groups. Primary group is made of fundamental institutions e.g., governmental, industrial, educational and family. Each of these primary institutions is associated with a subordinate secondary institution. The governmental institution involves a region of subordinate institutions such as constitution, legislation, political parties, civil service systems, tax regulations, tariff control and social security. In the same way religion involves several secondary institutions, such as beliefs, caste structure, rituals, symbols and taboos. In general an institution represents established ways of doing anything. This includes the working rules of the society under the control of economic, educational, family, legal political systems.

IMPORTANCE OF SPECIFIC INSTITUTIONS

The institutional factors which affect the ownership and use of land resources are sometimes described as landed institutions. In this way, the landed property right is the fundamental factor affecting land use, while some economic and non-economic factors also have greater impact in the ownership and land use. The government, law, customs, religion and education have also left their impact on judicious use of land resources.

Government and political institution : There is a considerable impact of these institutions on land use. This can be viewed from two aspects: (a) the overall impact of government policies and restrictions on public and private decisions regarding land resources, (b) the impact of organization and framework of government on the development and administration of public land resource policies.

Effects of Government policies and restrictions : Every decision of the government, especially that which is related to rural areas has affected the ownership and use of land resources e.g., levy can be used to force land into more intensive uses. Inheritance taxes can force the breaking up of landed estates and a vast ownership of private land.

According to the requirement of the society government may reserve large areas of public lands for parks, afforestation work, mineral exploitation, grazing and pasture land and for national parks and sanctuaries. For the use of military and defence purposes even private land may be acquired. Government has made several resource development programmes e.g. the construction of Bhakhra-Nangal dam in the Punjab and Himachal Pradesh, D.V.C. in Bihar and West Bengal Hoover dam, etc. This also includes the building of highways, the digging of canals, navigation and other facilities.

In the U.S.A. the State Governments have used their power to develop land use zoning ordinance, subdivision regulation, building codes, restriction in forest cutting, etc. as the guidelines for the private land use in the public interest. Other important regulation is for the provision and maintenance of parks and recreational areas, the location of building and streets and the management of public parking areas in urban centres.

Framework of Government : The organisation capacity and assignment of power at various levels, affect land resources according to their policies for land. The federal government can operate and administer its own land, set up and dispose policies and acquire private land for public use, etc. Government may provide fund for land use research, housing and reclamation projects, conservation of agricultural land, control of price, highway and canal construction scheme, programmes of agricultural development and the provision of land mortgage bank, cooperative bank, etc. These have both direct and indirect influences on land ownership and resource development. The problem of land use arises here in the border areas of two countries because no country can take independent view in the utilization of land in such areas. In Indian and other areas of the World State or local government, by their invested powers influence the land use to some extent.

Law and legal systems : These systems have their impact on land resources. The government is the framers of the constitution and in turn is governed by the set rules of the constitution. With this measure the government safeguards the claims made by a person or the society and favours property right on land against arbitrary action. Besides this, the government make laws, promulgate ordinance relating to the acquisition of lands, land registration Act and leasing and mortgage of land.

Custom and habits : These facts also have impact upon land use practice of an area. As for example in India a rich person often favours the planting of flowers in front of his house while a poor man often grows vegetables. Owing to food habits in India the food crops are highly raised while in the

U.S.A., besides this, large area is left for pasture land, as the population density is low and for rearing cattle wealth and poultry the land is sufficient there.

In the surrounding areas of an urban centre it is customary to invest land in market gardening, while in the rural areas generally food crops are grown.

A few decades ago cotton clothes were rarely available, while today besides this synthetic fibres are also used and hence this requires excess land to establish factories, etc. In the same way the increasing use of coal, oil, rubber, dietry items, writing papers, newspapers, wools, flex and nylon have variously changed the use of our land slowly, as our population is rapidly increasing. Property inheritance, land ownership are also affected by the custom.

Religious institutions : Affects land use in three ways—(a) ownership of land by religious institutions, (b) claims made by maths, temples, mosques, church and other religious organizations in the income of land and (c) religious beliefs affecting land use practices.

In historical times also particular sites were often regarded as holy spots. Today such sites are centres of religious repute e.g. Rameswaram, Deoghar, Rome, Mecca, Jerusalem and the Ganga river have become the local centres of pilgrimage. Graveyard and burial grounds have also been maintained for religious reasons and some quarters of productive land are retained for this purpose in many countries.

Educational institutions : The establishment of educational institutions, exert a considerable pressure on land, hence this affects the prevailing land use practices. Besides this, the specialized training of agricultural science, land tillage, crop rotation give an insight that induces the people not to depend upon traditional and customary way of agricultural practices and land utilization but to act according to their imagination. Thus technical abilities, resource development and urban planning have their direct influences on land use. The extensive use of mass communication media, e.g. public meetings, publication of newspapers and magazines, radio and television programmes, games and sports and film shows have increased the practice of new and improved land utilization techniques.

Other institutional factors : There are several other institutional factors which have considerable impact on land use. Family is an institution which provides incentive for settling down and carrying on its economic activities as commercial, industrial or agricultural. Among secondary factors transportation lines, banks and marketing facilities are important, which play a major role in determining the land use of particular area. These physical, economic and institutional factors control the success and failure of land use in any geographical region.

REFERENCES

1. V. Nath, Land utilization in India, *Journal of Social and Water Conservation in India*, Vol. I, No. 2, 1953, p. 4.
2. Paul Vidal de la Blache, *Principles of Human Geography*, London, 1959, p. 75.

3

Natural Principles of Land Use

INTRODUCTION

LAND USE of any place varies according to the variation in the spatial distribution of sun rays, rainfall, topography of the land, drainage condition, availability of minerals and the location of human habitations in relation to the market centres and transportation lines.

The spatial differentiation of these elements affect the purpose and character of land use. The whole land of the earth surface has, thus, potential value for some use or group of uses, although the area suited for any one use is often quite limited. So far as the principle of land use study is concerned, this can be studied under various uses e.g. agriculture including livestock farming, urban, industrial, recreational and transportational uses, etc.

GENERAL PRINCIPLES OF LAND UTILIZATION

The formation and functioning of land in relation to man is always governed by some inherent laws as the land has the productive capacity of foodgrains, minerals and water, and is the base of all human actions including shelter. In this way, the study of land utilization have some inherent principles which are as follows:

The principle of maximization: This principle emphasized that the utilizer of land and enterprises tend to maximise their value returns. The veracity of this principle lies in the assumption that economic behaviour is regional in the sense that people are motivated by a desire to make use of most of the resources at their command. Personal experience shows that people act in a manner approaching national behaviour because their actions are purely haphazard but reflect and attempt to maximise the utilization of land in an earnest possible manner. The practical application of this principle may also be illustrated by agricultural production in densely populated nations and in relatively new, sparsely settled areas. In a country where supply of labour is large and a huge amount of capital is accumulated through saving and thrift, only a small fraction of land is available for agricultural use. Agricultural production will be characterised by intensive cultivation with much input, capital and labour. Hence per acre yield will also be comparatively large. In contrast to this labour is relatively scarce in a newly sparsely settled areas of abundant agricultural land, but capital is even more scarce. Hence very few units of capital and labour will be applied in the given territory, and utilization will be extensive. On the other hand, large amount of land will be utilized in an extensive farming with only small investment of capital and labour per acre. In those areas the cultivated land operated by given individual tends to be several time more than that operated previously.

The use of land is also determined by the cost involved and the incurred profit. Thus the purpose of land use study is to search the possibilities of maximum profit after choosing the alternative better use. As for example suppose a farm is engaged in mango garden but the same is not profitable in terms of supplying food to larger number of people. Hence on such plots the cultivation of food crops with better infrastructure facilities could be suggested.

The equi marginal principle: The maximization of land utilization could be achieved through the operation of equi marginal principle. This can be stated after considering effective competition, the division of fixed quantity of use among the number of different uses so that they apportion each use an amount sufficient for increment of a gain and every use might be put for maximum profit. The transfer of resources from one use to another will occur until the rate of achievement from transference equals the rate of loss from withdrawal. A consumer will transfer his expenditure from food to clothing. In case the additional units of clothing will yield satisfaction equal to what had been gained from additional units of food, because additional food would have added greater satisfaction in comparison with clothing, and for this fact transfer of food into clothing on equi marginal principle is not feasible.

The principle of diminishing marginal rate of substitution: The increasing amount of production and consumption, the increment of satisfaction diminishing for any type of land utilization. It is true when two goods are substituted for one another. The smaller quantities of greater goods are necessary to compensate for the substitution of another good. This is a simplest form of the principle of diminishing marginal rate of substitution. As for example the use of land for vegetable cultivation is less and less in terms of production of rice as the farmer substitutes the first for the second. The same applies to an enterprise when machines are substituted for men. From the operational view of the principle the condition of equilibrium for the individual and also for the utilization of land as a whole is necessary. In case of determining marginal rate of substitution associated with the diminishing utility did not apply once a farmer began to use the land for vegetable cultivation.

The principle of diminishing marginal productivity : According to Malthus the growth of population surpassed food production and food supply and this principle is known as the principle of diminishing physical return. Malthus induced that land could not produce food as fast as population could expand. As our experience goes it is impossible to secure the same proportion of return with the additional input of labour capital and management given to a piece of land for indefinite period. The principle of diminishing productivity applies to the factors of production and determine maximum output with a combination of number of productive parameters.

The principles of specialization and comparative advantage: This principle of land utilization shows that each and every crop cannot be produced with the same amount of benefit and the same level of production. As for example, black soil is most suitable for cotton production. Calcareous mixed with alluvium and volcanic soils are suitable for sugarcane cultivation, etc. but any change in the above set pattern of specialization in land utilization may drastically reduce the production in comparison with cotton and sugarcane respectively. Hence in the selection of alternative use of land instead of cotton and sugarcane would result into loss of production and income.

The principle of first choice: The principle of comparative advantage is commonly referred as the principle of first choice. In case for the production of a crop only a limited number of land are qualified in production to the need, it will be the first choice of such lands. This principle can also be applied in terms of regional location of land utilization after considering the input-output ratio and the maximization of production, on a particular piece of land.

The principle of mapping and search for alternative better utilization of the land: Mapping the use of land through field to field survey, data analysis, cartographic analysis of maps in statistical analysis are the fundamental issue of this principle. This is because land is the base of all types of human decisions. Hence the search of alternative utilization of land involves the following issues:

—Land is the free gift of nature to man for sustenance of life either through forest gathering or subsistence agriculture,
—to increase the productive capacity of land,
—each land have fixed boundary according to human needs and physical resistance factor of nature,
—each land have owner under certain customary rights,
—each piece of land has some utility of agriculture, settlement road, fence, factories, playing ground, garden, aerodrome, etc.
—each piece of land used by man have some economic rent in terms of labour, mind, cash, distance covered and input,
—land use is the product of the demand set by people of any race, religion and income,
—hence with the concept of demand and supply under certain input-output ratio the competition between land use arises and this automatically pave the way for better alternative use of the land.

Optimization of the quality of land use in relation to environment: Under this principle it is impossible to establish a reindeer farm in Equatorial belt because one can rear reindeer in colder areas only, not in warm climate. Thus the optimization of the use of land should be searched in the similar environment with favourable ecological conditions and in the same environment in which maximum production could be achieved with better infrastructure facilities.

Principle of increasing the efficiency and capability of the land: This principle is to increase the efficiency and capability and capability of land after knowing the shortcoming through land use survey and classification. With the help of better irrigation, seeds, fertilizer, deep harrowing and intensive cultivation the efficiency and the capability of land could be raised.

Fixation of cropping system around human habitation for least effort and less cost travel: Man tries to set things at accessible points with less cost, and time involved in travelling. Hence the desire to see the presence of almost all categories of land use give rise to competition in between the varying objectives of the use of land. This is because the perishable and costlier vegetables and crops are generally raised near the settlement, while less costlier crops find places away from the village site. For the presence of inherent quality, fertility, productivity of the land, soil conservation practices, manuring, management of irrigation, etc. are important aspects which should be essentially considered by a land use investigator.

Conservation of land for future judicious use: Conservation means the reduction of the rate of exhaustion of productive capacity of land resource in order to bring more and more land under cultivation. This includes the saving of land from the hazards of erosion, insects and pests and on the other hand, the input of fertilizer, irrigation, intensive labour and defence of crops. This also includes consolidation, improving the techniques of farming and proper judicious use of land by avoiding wasteful cultivation.

AGRICULTURAL USES

For agricultural uses the sunshine should be sufficient in most areas as the low temperature prevents all types of agricultural uses. O. E. Baker has estimated that about one-fourth of the earth is cold and only forest can survive there, i. e., in northren Alaska, Canada, Scandinavia and Soviet Union mainly forage for wild life, sheep, cattle and reindeer can be raised. On the other hand, in the cold deserts of arctic region, the frozen ground is unsuitable for any kind of agricultural use.

The availability of water has also varied effects on agricultural land use. In temperate regions only 37 centimeters of rainfall is sufficient for agricultural land use, while on an average 100 centimeters of annual rainfall is necessary in tropical areas. Only 70 per cent of Europe and South America, [illegible] per cent of North America, 29 per cent of Asia, 25 per cent of Africa and 9 per cent of Australia receive adequate and reliable amount of rainfall. On the earth surface only 34 per cent area has availability of sufficient amount of water for agriculture and 1 per cent i. e., 200 million acres is benefited from irrigation. In this way most of the world's wasteland still await development for agricultural use. Only the wet land is suitable for growing rice, jute and other water loving crops, while in areas having dry climatic condition with low rainfall, dry farming, pasture and grazing land can be developed.

Even in climatically suitable areas of mountainous and rugged terrain, generally the cultivation of land is not possible. There only terraced cultivation can be successfully practised. Pearson estimates that 'only 64 per cent of the world land is topographically favourable for crop use'. In mountainous region, the terrace cultivation requires huge capital, time, human effort to tame the adverse nature and hence these prohibitive factors give rise to forestry and grazing ground.

The character of soil, such as its colour, texture, structure and chemical properties determine the land use of a region. They vary from light coloured red to black soil, which ranges from heavy clay, sand, gravel to alkaline. As Pearson remarked that 48 per cent earth surface is covered with productive soil for good agricultural use. But these estimates of soil and crop land use involves value judgment, which may vary according to personal knowledge and experience of the investigator. The deficiencies of soil may also be induced with the application of fertilizer, irrigation and other known methods of land conservation.

NATURAL LIMITS OF URBAN LAND USE

Locational factors have differential effects on urban land use and the rise and fall of cities. Urban centres are the result of the productivity and the size of hinterland. The intensity of land use around an urban centre in the determinant factor of its location and future extension. Towns and cities are, thus, found near the productive land, harbour sites, navigable streams, rail heads, road side and other spots favourable for trade and commerce.

The climatic control, conditions of, soil and productive ground plan are not needed for urban site. On the other hand the intensity of urban land use is determined by defensible higher site above flood level, social cohesion, functional specialization and population size of the settlements. Generally, higher site settlements and planned colonies bear no drainage problem but in unplanned colonies congestion and drainage problems give rise to slum conditions.

The favourable condition of top soil is necessary for successful operation of market gardening and growing of flowers. Even if the original soil is unfertile, the soil brought from outside can add to the fertility. Topographic condition is of considerable importance in determining the land use of any area. Usually higher mountainous land is not suitable for cities growth but the summer resorts and hill stations may develop as cities due to the attraction of large number of travellers and natural sceneries. Simla, Nainital, Mount Abu and Ooty arc good examples.

NATURAL LIMITS OF RECREATIONAL LAND USE

Sites of scenic beauty, historical significance, health resorts are important facets of recreational land use. Under scenic beauty higher importance is given to the deserts, dunes, underground caverns in limestone region as in Yugoslavia, mountain lakes as in Kashmir, national parks and sanctuaries as in Gir forests, fishing ground, hunting of wild life, ocean beaches and others. Intensive recreational land use is found in urban areas, which are endowed with the theatre ground, exhibitions, cinema houses, play grounds and gymnasiums. Similarly a variety of flora and fauna, water sites, swimming pools, horse riding, boating and other recreational activities find place in urban land use.

TRANSPORTATION, SERVICE AREAS AND MINING USES

Transportation routes are life lines to determine the land use. The location of air ports, rail heads and road networks facilitate the commercial and communication activities. The presence of intervening monticles, water-bodies and snow cover create hinderance to transportational linkages to some extent. Public buildings, cremation ground and the water well must be located at easily accessible point to the whole community.

Under service areas, the military cantonments may reserve sizeable portion of land for firing range, training grounds and residential quarters. The reservoir sites and watershed areas may be used for multiple land uses, in case the surrounding areas are susceptible to flood and unfavourable terrain conditions.

In comparison with all other land use, the areas engaged in mineral excavations are less dependent upon favourable location i.e. climate, topographic condition of land and soil types. Owing to economic gains of minerals, their excavations take place from far mook corners of the world. For these reasons oil-wells and mining camps are always located away from the industrial sites. This is possible with the help of accessible highways and development of energy resources.

On the earth surface 72 per cent of its area is covered by the Ocean, 18 per cent by the desert, mountain and polar areas and only 10 per cent have suitable ecological conditions of climate, topography and soil necessary for the production of food. With the help of the above background, it is now possible to discuss the principles of land use.

FIRST PRINCIPLE OF LAND USE

The area of earth surface is fixed and in no way it can be extended like an elastic material. One may add a fraction of land after reclaiming the coastal strips, marshy ground and stony waste by the conservation practices. The population is increasing rapidly in the world. This shows that per capita land is decreasing day by day. In this way a day will come when man will find only a standing space over the earth surface. One should always think about this, while putting the land into various uses and planning the land for better uses.

SECOND PRINCIPLE OF LAND USE

Broadly, the physical characteristics of land are almost permanent as plains, plateaus and mountains. Not a single physical feature can be transformed into another, except by large scale diastrophism. Hence in spite of the super knowledge of man, one has to be satisfied with the fixed piece of land as provided by the nature. Although for our satisfaction, we are still engaged in reclaiming the hill slopes, waterlogged areas, coastal stretches of lowlands, even if these are limited gains for our community. To harness the utmost capabilities of the land through scientific methods of cultivation, fertilization and irrigation, the inherent original capabilities of land may be upgraded but it is an expensive process. The areas abandoned by mining operations should be levelled and cultivated with great care, so that the restoration of its best quality may be maintained, even if its economic importance is insignificant. Thus one should recognise the influence of physical form of the land and its best form of utilization.

THIRD PRINCIPLE OF LAND USE

Land use is the product of ecological balance of various parameters, existing over any space (Fig. 3.1). With the casual relationship of these factors, the land use vary from place to place, according to the changing condition of various natural factors. The arctic region is sparsely populated due to cold weather, although the other factors are the same as in other parts of the globe and hence even if one factor is unfavourable it may change the use of the land. In desert regions the scanty rainfall and the sun-baked soil retards agricultural practices and in such environment people led a nomadic life. The cultivation here is found only on oases with the provision of irrigation. In comparison with the desert, alluvial soil is supported by the annual flooding which replenished its fertility to sustain the heavy pressure of population since ancient times. Temperate woodland supports subsistence farming whereas the grass lands are the graineries of the world. But the increasing demand for land by the growing population and the reckless cutting of forests and animal grazing has led to soil erosion which is also creating problems of land conservation. Thus one should restore and maintain the ecological balance by the way of land conservation practices.

FOURTH PRINCIPLE OF LAND USE

The land use of a region is the product of its energic system and production potential. Since prehistoric time, owing to the excessive harnessing of the land resource, the natural fertility of the land is considerably reduced. The complexities of human society have increased the intensive use of land, especially in China, India and Holland where the pressure of population on land is higher in comparison with the less populated and the extensive farming areas of the U.S.A.

Each piece of land has inherent characteristics of capabilities that determine the value of land for any required use. Thus, it is subject to the law of diminishing return. The productivity of the land is not a static force, but, it is subject to time, space, cultural attainments of the society and natural influences of temperature, pressure, rainfall and the presence of essential ingredients of the soil. In a certain level of cultural attainments, the society acts with varying intensity within a bond of economic, social and technological advancement which affect the utilization of the land paramountly. This shows that cultural environment is inseparable from man and land use in the expression of natural as well as the cultural environment which are inseparably related to each other.

In recent years, the immense technological advancement has challenged the natural limitation of the land and the claim is made that modern science can grow food up to any quantity for the consumption of human society. But our history shows that natural environment has always been a powerful force in determining the land use and the concentration of population over a piece of land.

FIFTH PRINCIPLE OF LAND USE

Land use possesses human value in addition to its production potential. All the cultural elements of landscape are the products of land use in the forms of human dwellings, factories, monuments, tanks, roads, railways and air ports which are located according to the differential needs and preferences of the people. All these are located on land and have special value to satisfy the increasing human needs, besides fulfilling our desire for food, cloth and shelter. The loss of agricultural land at the cost of urban expansion, transport development and growth of utility services are the vital problems of land use today.

The intangible value of land use indicates that "man does not live by bread alone". The small working hours and longer absence from duty demand more opportunity for outside enjoyment in parks and play-grounds which require the increase of recreational areas of land use in urban centres. On marginal lands of forests and plateaus the land is devoted to conserving our vanishing wild life in national parks and sanctuaries. In Chota Nagpur Plateau of South Bihar, the Hazaribagh and Betla national parks and Adhaura sanctuary of Rohtas district are the best examples of judging the values of such land use to maintain the ecological balance as wild life eventually supports our protein supply. Their scenic value, grandeur recreational use and the supply of raw materials for our industries like bidi, silk, lac, fruit tanning, lumbering, paper and several others are the best examples of human values of land in addition to their production potential.

4

Approaches of Land Use Studies

THE PROBLEM of food availability and increasing the standard of living of the masses has intimate relationship with the land utilization. The scientific expression of real world situations and their generalizations have been delineated here in terms of approaches of land use.

The approaches of geographical investigation of land use study are viz., ecological, plot to plot land use survey and the quantitative techniques. Quantitative method yield direct information on factor analysis which is adopted for a better understanding of agricultural land use. More emphasis is given on ecological approach as to how the association of environmental factors control a particular farming system. It is also helpful in determining the natural boundaries of land use categories conditioned by farmers behavioural approach on plot to plot basis.

Recently the approaches of land use study have changed from descriptive to cause and effect relationship analysis followed by several other approaches which are as follows:

Genetic approach: The study about the genesis of land use pattern is known as genetic approach. Primarily in forest environment no one cares to study the use of land. But today with the increasing demand set on a particular piece of land, the forest land is now used to serve several odd functions. Thus the circumstances under which a land is set to perform the set task comes under genetic approach. As for example in a pioneer fringe the land is covered by dense forest. But with the migration of population and rise of settlements in those areas will automatically set the land to perform the task of agricultural land, roads, culturable waste land and others.

In this way the genesis of each and every plot of land can be estimated after considering the historical evolution of the cultural elements of the landscape.

Prescriptive approach: Land use study is directed towards the better utilization of limited land for the welfare of human society. In this sense the prescription about the use of fertilizer, reclamation of forest land directive for judicious future use of the land, etc. can be suggested by scientists. Thus the prescriptive approach is the best for planning the future use of the land.

Remedial approach: Under this approach the investigator is to search the alternative better use of the land by suggesting the removal of a certain inferior use. As for example the conversion of marshy ground into agricultural use, the planning of settlement on unproductive, defensible higher site instead of good agricultural land, etc.

Case study approach: Under this approach the detailed analysis of the use of every piece of land in a village, anchal and district or part thereof are included. This study will be contemplated after random, purposive, stratified and multi-stage sampling.

Regional approach: Under this approach, the influence of the regional geographical factors on the distribution of crops, fields and other use of the land are included after considering the region as a whole.

Eco-system approach: This approach deals with the association of environmental factors which affect the location of crops or any other agricultural operations and vice versa through feedback mechanism process. This approach focuses attention on the basic properties common to all systems: structure, function, equilibrium and change. It leads to ask five fundamental questions about each land utilization system: How is it organised or what is its structure? How does it function ? What degree of stability does it have? How did it evolve through time ? How will it develop in future?

Statistical approach: In this approach with the help of land utilization and agricultural statistics some unseen inferences can be obtained by the analysis of agricultural situation of a region.

Principle approach: Under this approach Von Thunen model of land use explanation through the formation of ring around a settlement, Malthusian approach and Marxian approach have been analyzed after considering the pattern of land use of a region under study.

After considering the above the following workable approaches for the study land utilization are as follows;

(i) origin and growth of the methodology of land use study,
(ii) relationship with economic, agronomy, physical geography, climatic conditions and other social sciences should be considered,
(iii) field survey, first hand observation and collection of data,
(iv) knowledge of season and crop report, maps, economic tables and documents, etc.,
(v) data analysis, computing, map processing, facit handling and computer programming,
(vi) conceptual framework of agricultural regions, land use intensity, location, form, structure, distribution and productivity of crops, land classification, innovation and diffusion of seeds, fertilizer farm technology, including the study of agricultural efficiency, crop combination, land conservation and planning the land use, etc.,
(vii) characteristics and elements of agricultural practices and interrelationship among them i.e. theoretical, organic as well as eco-system and behavioural approach,
(viii) development of hypothesis and models, etc., and
(ix) Empirical and statistical analysis, etc.

THE MALTHUSIAN APPROACH

This is an approach to study the problems of land use. A book named *Road to Survival*[1] deals with the Malthusian approach along with the book *Readings in Land Utilization*[2] There is still another book *Food Planning for Four Hundred Millions*[3] which also includes the Malthusian approach to the problems of land utilization.

The Malthusian concept of "struggle for land" shows the importance of land as a factor of production in relation to the evolution of human settlements and degrees of land utilization. The original idea of Malthus was on population checks. He felt that the overpopulation leads to overburden on every piece of land but it is always countercheked by nature. He had shown that the economic effects of over population are uneconomic or undersized holdings which prevent the adoption of improved methods of cultivation. This leads to the abolition of following and enhancement of soil erosion with deforestation especially on fertile plains and marginal lands. In this way the general costs of cultivation are raised and the proportionate labour done by men and bullocks is increased. The net result is low yield per unit of labour, agricultural idleness and unemployment are linked with one another creating a vicious circle. The mobility of population and the combination of small farming with agricultural labourer, small implements are mitigating the overcrowding in agriculture. The lower rate of industrial growth is bound to fail in drawing off the surplus population from the land adequately. Thus owing to the increase of a landless class, it is inevitable that the overcrowding in agriculture cheapens the labour in comparison with the availability of capital and this may also delay the introduction of improved agricultural tools and intensity of land use.

Further, Malthus had shown the impact of drought, famine and epidemics like cholera, malaria, etc. as mortality and low birth rate of child and how these are closely related to malnutrition and low caloric intake. It is also felt that in the absence of agricultural surplus in comparison with increased population, malnutrition affects children and women indirectly which leads to shrinkage in the birth rate with spatial variation in age, sex and racial structure.

There is a variation in birth and death rate with respect to harvest conditions of crop in many parts of India, indicates not only the low agricultural return but also the vulnerability of population due to high population density. In Malthusian checks the productive power is restricted in case of lower economic strata by malnutrition or deprivation of certain essential elements e.g. calcium and vitamins of wheat, which is superceded by millets, milk and even non-availability of vegetables on land in unfavourable years. The malnutrition affects physiological vigour which leads to an increase of abortion, lower fecundity and infantile or maternal mortality. Hence climate, race, social customs and economic conditions have differential effects in controlling the growth of population besides natural checks, as the pressure of population on land has developed a varying intensity of land use in different parts of the globe.

The book *Road to Survival* leads to the following idea of Malthusian approach:

(a) The nutritive qualities of land help in producing varying qualities of vegetation and hence, the productive capacity of individual farms, nations and continents varies from one another. It is felt that drought in India caused repeated failure of crops, and with the incoming of famine it sharply affects the mortality of human being and livestock. Thus the carriers and barriers of favourable and unfavourable conditions of environment on the earth are the guiding factors of global productivity and supporting capacity of human being. It has also realized that environmental resistance has effective power of limiting land use and wherever its affects decreased, the productivity and the carrying capacity of the land rises. The carrying capacity if the land may be defined as the resultant ratio between the productive capacity and the environmental resistance.

(b) The lust for longevity of life is made easy by medical facilities in comparison with that which was 2000 years ago. But according to Vogt "medical care and improved sanitation are responsible for more millions living more years in increasing the misery".[4]

(c) Vogt discarded the idea that industrialization alone can produce high living standards. He ranked countries in descending order of their living standards viz., before Second World War Canada, New Zealand, Argentina and Australia were highly industrialized countries with lesser population in relation to their land and other natural resources. On the other hand, there are countries which have higher population and higher industrialization, i.e., Sweden, Germany, Czechoslovakia, Japan and Italy which were not ranked low in standard of living in comparison with the above-mentioned nations. He further claimed that Indians should be allowed to emigrate to other under populated countries e.g., Australia, Canada and the U.S.A., otherwise the turmoil will stultify industrial development and Malthusian checks will hold the population under reasonable bounds as it was in pre-British Regime.

(d) It was felt that only the available resources of the country, the present trends of carrying capacity of the land and per capita caloric intake can decide whether any adjustment of excess population is possible or not.

(e) After the collection of information for per capita calories, etc., Vogt suggested to launch comprehensive conservation programmes together with the measures of family planning in each and every country. As according to 'Food and Agriculture Organization' the areas of overpopulation should extend food production programmes along with the contraception programmes, but keeping in view the Malthusian approach one "should not ship food to keep alive ten million Indians and Chinese this year so that fifty millions may die five years hence".[5]

THE MARXIAN APPROACH

This approach of land utilization is devised by Lenin and Stalin. They feel that with the passing away of time the diminishing return of the productivity of land and overpopulation are wrong concepts of Malthusian approach.

In his celebrated book *Theory of the Agrarian Question* Lenin presented the following facts against Malthusian approach:

(a) About the 'law of a diminishing returns' it is said that "if each additional investment of labour and capital in land, produce not a diminishing but an equal quantity of products, there would be no sense in extending the area of land under cultivation, additional quantities of grain would be produced on the same plot of land, however small, and it would be possible to carry on agriculture of the whole globe upon one desyatin of land".[6] This idea on law of diminishing return has lost sight of the level of technical ability, the state of productive forces and the additional investment of 'labour and capital' which predetermine the dynamic method of production and reforms in technique. In increasing the capital investment on land up to considerable degree, the invention of machinery, new system of land cultivation, new methods of livestock farming and transportational goods are required but the additional investment of capital and labour may take place in case the technique of production remains unchanged. Thus in changing methods of production, the law of diminishing return does no longer hold good.

(b) In Malthusian approach it is said that "the development of civilization and increase of population, food production is becoming more and more difficult".[7] Regarding this Lenin[8] expressed that it is silly to think that primitive men obtained all their requirements free from the nature. In reality primitive men had problems of existence and difficulties of fighting against nature. In conditions of increaseing population, it is not difficult to produce food but it is difficult to obtain it for the workers, as the capitalistic society has concentrated the land, machinery and money in fewer hands which restricts a successful production and the best utilization of land.

(c) Lenin completely rejected the Malthusian concept of 'overpopulation' and in this connection Bulgakov's statement is as follows: the areas of overpopulation may be regarded as 'absolute poverty' areas which faces the problem of meagre production and not of distribution.

(d) In Malthusian approach the fact that "economic progress and historical development is determined by geographical environment" is rejected by the Marxists.[9] At this point, the view of Stalin is as follows:

> Geographical environment is indispensable for the development of society but it may not be a determining factor. In India since prehistoric period three different social systems have been successively superceded, the primitive communal system, the religio-ritual system and the modern system. In these periods, the geographical conditions were more or less the same in India with slight variations. It is, thus, a fact that geographical environment may not be the determining factor of social development and economic progress of a society.

(e) It is also said that population growth is the determinant of social systems. But the Marxists ask as to why the primitive communal system is succeeded by the religio-ritual system and the latter by the modern system and not by others. The reply would be that the determining factors are the methods of procuring livelihood, the instruments of productions and human abilities including experience and skill for boosting up the productive forces of the society.

L.A. SALTER APPROACH

Salter approached the problems of land utilization with greater emphasis on the functional use of hypothesis in his book, *A Critical Review of Research in Land Economics.*

1. In future the need is to apply social science research in the development of land utilization in a society. The rearch should be such in which theory can be applied into practice. It means that research is not only a method of collection of data and technique of observing anything minutely but it has the ability to convert theory into practice.

2. The research on land utilization should be applied and problem oriented. In search of the solution of any problem it is highly needed, because research is the central point in between theory and practice. These are the basic problems of rural land economics especially in the use of new fallow land. The development of landed property is closely related with the people configuration of the land and intensity of demand as set by people.
3. Salter had stressed the practical utility of any hypothesis. This is because hypothesis gives alternative line of thinking in fulfilling the desired goal. The function of investigator is to collect data for guiding the research. The purpose of research is not only to accept or reject hypothesis but its purpose is to expand and modify the result till it gives desired result as our experience goes. It reveals that which action will give satisfactory result of land utilization. This will also provide desired landed property relations. It is also necessary in order to overcome the changing conditions of land utilization and to convert theory into practice.

In this way in order to solve the problems of land utilization there are three processes which are clearly revealed in the works of Vogt, Lenin and Salter. The theory of Malthus which Vogt has presented shows the patterns of relation between man and the land the kind of social tensions it arises. On the other hand, the theory of Marx had completely rejected the theory of Malthus keeping in view the changing conditions of society and geographical determinism. But it shows relations between man and the productivity of land. In order to solve the problems of land utilization Salter has led greater emphasis on the functional use of hypothesis.

URBAN LAND USE A DEVIATION FROM RURAL LAND USE

Urban land use deals with the problems emerging in urban cèntres in the process of selecting and translating into action, the optimum utilization of limited land between shopping centres and residential areas.

The geographical distribution of land use in urban areas is of prime interest to the geographer, urban sociologist and the City Planners. The differentiation in land use pattern in urban areas vary on account of internal differentiation in the location of commercial establishments, administrative offices, factories and residences.[10] Owing to rapid increase in population and size of the urban areas, land use changes add a dynamic character to urban land use in comparison with the rural sector. Urban geographers in India have also contributed to urban land use study e.g. Gist (1958),[11] Rajagopalan (1960),[12] Chatterjee (1965),[13] Alam (1965)[14] and Singh (1966).[15] This section deals with the basic principles of urban land use study in relation to rural land use.

An urban centre is a collection of house of non-agricultural people. Here various uses of land are found on account of, (a) the forces of attraction and integration, (b) the forces of dispersion and disintegration, and (c) the forces of spatial differentiation (Dickinson, 1964, 125).[16] The first two help in the vertical growth and horizontal expansion while the third results into segregation of buildings, persons and activities. A little change in social and economic conditions caused by natural or political forces, brings a remarkable change in land use patterns. A plan through land use study to eliminate some of the undesirable changes and to minimise the problems that have been already posing becomes imperative. In guiding the urban activities, the pattern of land utilization is of paramount importance, hence a clearcut land policy becomes essential for guiding the use of the land, curbing the misuses, preventing the abuses, regulating the non-uses and aiming at the reuses of land in the better interest of the inhabitants (Abrams, 1953, 34).[17] The study of existing land use is essential before starting planning the urban areas.

Land use planning in urban areas is a part of a larger process of city planning. It is 'basically concerned with location, intensity and amount of land development required for various uses of space, functioning of city life, industry, wholesaling, business, housing, recreation, education, religious and cultural activities of the people' (Chapin, 1957, xiv).[18] Land use is determined by land economics, social values and public interests. In urban areas certain sites have greater value for certain specific purposes. This value is established by a web of relationships with other surrounding uses and their value structure. Social values and human behaviour play a great role in the evolution of land use patterns, through such socially rooted determinants are less understood and are frequently confused with economic determinants. The urban land use study is a new branch of geography and its importance has been realized since the Second World War when the need for planning and replanning of urban places was universally felt. (Dube, 1969, 26-34).[19] Land use mapping in planned urban centres are easier in comparison with old and complex urban areas.

Land use varies according to the differential demand and requirements of the people over a piece of land. For this fact, the land use pattern, as well as the amount of land utilized for a particular purpose, and often the intensity of development are constantly under change, according to the changing needs of the community.

URBAN LAND USE CLASSIFICATION

The continuity of land use always presents a problem in classification. Urban land use and rural land use are used to describe activity systems for which there is general agreement on interpretations. But it is not known that where urban end and rural begin as one travels out from the city. Rural-urban fringe is a transition zone where the remnants of agriculture persists besides residential quarters, air ports, power plants, electrical sub-stations and similar odd functions.

There is no agreement upon uniform classification of urban land uses, adopted by individuals and organizations in different parts of the world. This is primarily due to differential land uses in urban areas, and difficulties in mapping various uses. It also varies because of differences in the purpose for which they are mapped. There is no spatial relationship between land uses of Indian urban areas and those of the Western Countries. The highly mixed land use in Indian urban areas differs markedly from the usual segregation of land uses in Anglo-American cities (Breese, 1969, 58).[20] It appears to be the result of the long history of growing compactness and continuing necessity of travelling between places of residence and that of employment.

There are four major categories of land use in urban areas, i.e., agricultural, residential, commercial and industrial Gallion and Eisner, (1965, 191).[21] Bartholomew (1955),[22] distinguished two categories of urban land use areas: developed and undeveloped (vacant). Then he divided the developed areas into (a) privately developed, such as residential, commercial and industrial and (b) Publicly developed, such as streets, transport establishments, parks, playgrounds and public and semi-public space and public buildings have been recognised as major land use categories, besides others (Berry and Horton, 1970, 444-49).[23] Chapin (207) categorises urban land use into residential, retail business, transport and communication, industrial, wholesale and storage, public buildings and open spaces and vacant or non-urban use. There were six land use categories in England, in 1950s: residential, central areas (administrative and commercial), industrial, educational, open land, large establishments, hospitals and cantonments (Housing and Local Government, 1952, 37).[24] Collins (1965, 217-29)[25] has classified the categories of land in Britain as residential, open space, public buldings and institutions, industry, commerce, statutory undertakings, vacant and derelic buildings, while Conzen (1966, 56-78)[26] classified the urban use of Britain into following categories:

1. Shops, business and professional premises with or without residences.
2. Premises with Public or Community Functions.
3. Public Transport Premises.
4. Industrial Premises including Builder's Yards and warehouses.
5. Residential Premises.
6. Open Spaces.

In America, public buildings are public properties, while in Britain, public buildings are not essentially publicly owned. Even a privately owned house may be called a public building if it is used by the Government for various official purposes. A land use map therefore, is not a map of land ownership or land occupancy (Jackson, 1968,113).[27]

In India, land use categories recognised by different scholars belong to two different types, e.g. town planners and urban geographers. Town planners quite often categorise urban land use as residential, commercial, industrial, transport, communication, public utilities, open space, agricultural, vacant land and water bodies (Gowda, 1972, 86-87).[28] On the other hand, there are minor difference amongst the urban geographers and most of them classified the urban land use into the following categories: residential, agricultural, open spaces, military lands, parks and burial grounds (Singh, 1964,[29] Sinha, 1970,[30] Alam, 1972,[31] Ram and Thakur, 1972).[32]

The lack of uniformity in land use categories is due to the absence of detailed data for various land use categories; non-uniform distribution of land under different categories and difference in the character of individual cities and experience of the investigator. This difference is lesser in major categories while it is higher in minor categories, as religious, recreational, medical, etc.

In India, based on land use classification of planning organizations and urban geographers, J.P. Singh (1976)[33] prepared the following list of land use categories for the study of Shillong Municipal Area in North Eastern India.

Residential (R): Boarding houses, clubs, welfare houses, residential quarters and hostels.

Commercial (C): Hotels and eating places, stores, petrol pumps and service stations, autorepairs, shops, bus transport depots, big commercial agencies, cinema and banks.

Industrial (I): Factories, saw mills, autobody builders and iron and steel fabricators.

Agricultural (AG): Vegetable and fruit gardens and farms producing food grains and cash crops.

Educational (E): Schools, colleges, training centres, universities and libraries.

Medical (M): Public dispensaries, hospitals, family planning centres, health and social centres and veterinary hospitals.

Administrative (A): Government offices, post offices, police stations and fire brigade.

Defence (D): Cantonment, various defence camps, police lines and rifle range.

Forest (F): Protected, reserved and other forests.

Places of Worship (W): Churches, temples, mosques and Sunday schools.

Open Space (OP): Gardens, parks, race course, golf links, sports grounds and vacant lands,

Graveyards (GR): Burial grounds.

CONCEPTS OF URBAN LAND USE

Many scholars like Burgess, Ullman, Hurd, Hoyt, Feiry, McKenzie and Colby have presented their concept of urban land use models to express city structures and to generalize the arrangement of land use regions within a city. These models have been affected by physical, economic, political and cultural factors of the urban areas in shaping the form and structure of cities over decades. These variables are interlinked with the process of feed-back mechanism in the repetitive land use patterns in all urban centres. Four well known models are as follows:

Concentric zone theory : A hypothetical model of urban growth (Fig. 4.1) was developed by E.W. Burgess, a Chicago Sociologist in 1925.[34] This model suggests that any urban area tends to expand radially from its central commercial core to the outer periphery, so as to form a series of concentric circular zones which are as follows:

The central business district : It has been considered as the focus of commercial, social, civic life and transportation. This is the commercial hub with its down town retail districts, stores, shops, office buildings, clubs, banks, hotels, theatres, museums and organisational headquarters. This zone of retail business is encircled by wholesaling.

The zone in transition : It is known as the areas of residential deterioration where business and light manufacturing encroach upon residential areas. Here slum prevails with their scattered pockets of poverty, degradation, diseases and underworlds of vice. In American as well as in Indian cities, it has been observed that recent immigrants inhabited these regions because of lower house rent and easy access to the business centre.

The zone of independent workingmen's homes : In this zone

mainly industrial workers live in American cities, who have desire to live within easy access of their work.

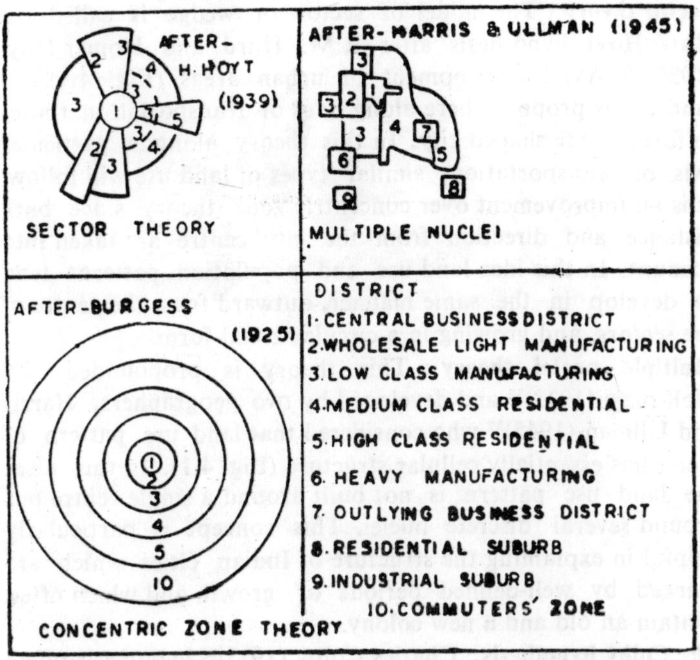

FIG. 4.1 : *Patterns of Land Use in Urban Areas*

The zone of better residences : This is known as restricted areas of high-class buildings, where mainly single family resides.

The commuters zone : In areas of rural-urban fringe, the spotty development of high class residences along lines of travel is known as commuters zone.

In these inner-structure of the urban land use it has been noted that each inner-zone has tendency to extend its area by the invasion of the next outer zone with the process of second zone of transition. According to Burgess, this dynamic process of city structure, shifts sorts and relocated individuals and groups by residences and occupation create cultural and economic differentiation among various groups of people and land use of the urban centre. It has been further noted that land use pattern outside the central business zone varies with some gradients which may be discernible in delinquency rate, sex-ratio, percentage of foreign born individuals and poverty tending to decrease outward from the central business district.

Sector theory : The model of sector or wedge is called the Hurd-Hoyt hypothesis after R.M. Hurd and Homer Hoyt (1939).[35] Axial development of urban areas (Fig. 4.1) and land use is proposed here along lines of transportation routes to form a star shaped city. In this theory along a particular axis of transportation, similar types of land use will follow. It is an improvement over concentric zone theory since both distance and direction from the city centre are taken into account. In this idea land use and population patterns tend to develop in the same manner, outward from the centre of the sectors, and growing in a circular zonal form.

Multiple nuclei theory : This theory is propounded by McKenzie (1933)[36] and developed by two geographers, Harris and Ullman (1945)[37] who considered that land use pattern of a city has essentially cellular structure (Fig. 4.1). In this idea, the land use pattern is not built around a single centre but around several discrete nuclei. This concept is particularly helpful in explaining the structure of Indian cities which are marked by well-defined periods of growth and which often contain an old and a new colony.

The Colby hypothesis : Charles Colby (1933)[38] hypothesis offers a dynamic explanation of urban development. In this theory, the pattern of land use in any city at any given moment is the result of two forces i.e., centripetal and centrifugal. Centripetal forces are of two types:

(a) Residents and business communities rush towards city attracted by the service amenities, employment opportunities and other economic advantages in an urban area, (b) within urban areas itself, both people and businesses on the fringe are drawn towards the C.B.D. because the central point provides the least access to either customers or labour or other institutions such as stores, theatres, libraries. On the other hand, centrifugal forces drive people and business out of the C.B.D. into peripheral areas. This is due to the flight from congestion and longing desire for open space.

These theories of urban land use are not mutually exclusive (Nelson, 1969, 201).[39] The concentric zone theory depends upon the concept of centripetal and centrifugal forces as envisaged by Colby, while the sector theory and multiple nuclei are modifications of the concentric zone theory. In Hoyt's concept, residential area expand outwards concentrically and Harris and Ullman's theory is basically the same as that of Burgess, so far as a nucleus is concerned.

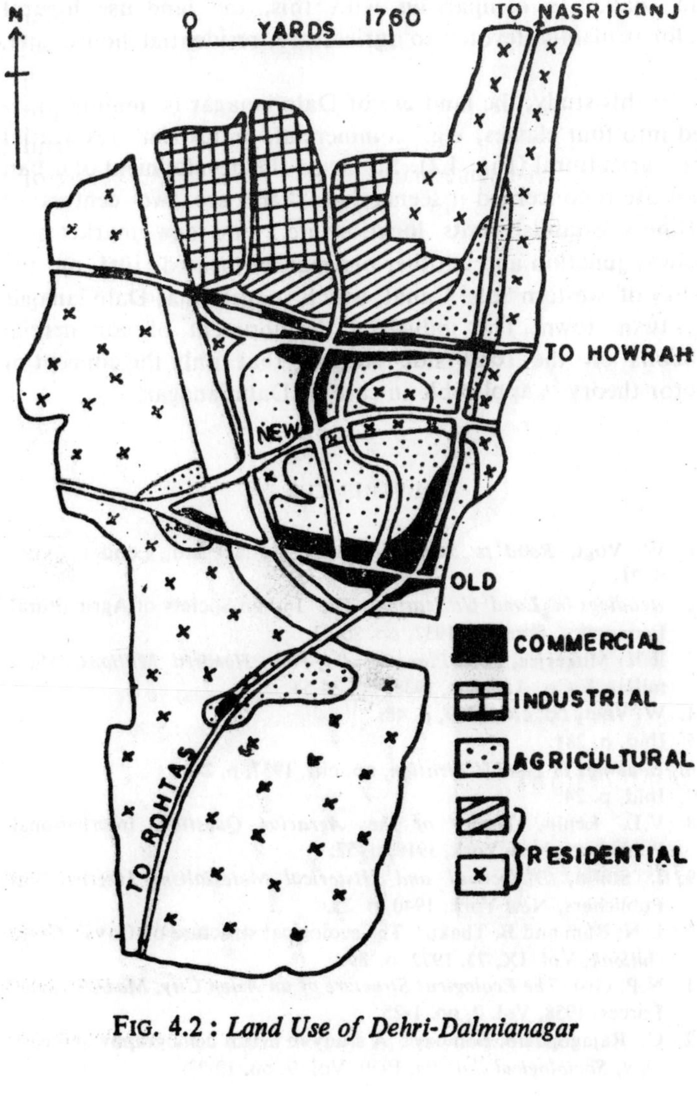

FIG. 4.2 : *Land Use of Dehri-Dalmianagar*

DALMIANAGAR: A CASE STUDY OF URBAN LAND USE PATTERN

The development of Dalmianagar as an urban centre in Rohtas district of Bihar is due to the connection of Grand-Chord line, Dehri-Rohtas light railway link and the development of number of industries e.g. paper, sugar, cement, hydrogenated oil, heavy chemical, vulcanised fibre plant, asbestos, cement and soap.[40] In comparison with this, the land use in rural sector is mainly devoted to agriculture, residential house site, etc.

In this study, the land use of Dalmianagar is mainly divided into four classes, e.g. commercial, industrial, residential and agricultural (Fig. 4.2). So far as the development of urban land use is concerned it seems that there are two centres of business establishments location i.e., one new market near railway junction and another old market located just on the bank of western Sone Canal, which suggest that Dalmianagar is a twin town. The industrial development of commercial ribbons on the road side suggest that only the concept of sector theory is applicable in case of Dalmianagar.

REFERENCES

1. W. Vogt, *Road to Survival,* Victor Gollacz Ltd., London, 1949, p. 21.
2. *Readings in Land Utilization,* The Indian Society of Agricultural Economics, Bombay, 1957, pp. 20-23.
3. R.K. Mukerjee, *Food Planning for Four Hundred Millions,* Macmillan & Co., London, 1938, pp. 34-54.
4. W. Vogt, op. cit., 1949, p. 48.
5. Ibid, p. 281.
6. *Readings in Land Utilization,* op. cit., 1957, p. 24.
7. Ibid, p. 24.
8. V.L. Lenin, *Theory of the Agrarian Question,* International Publishers, New York, 1918, p. 52.
9. J. Stalin, *Dialectical and Historical Materialism,* International Publishers, New York, 1940, p. 25.
10. L.N. Ram and B. Thakur, The ecological structure of Gaya, *Geog. Outlook*, Vol. IX, 73, 1972, p. 89.
11. N.P. Gist, *The Ecological Structure of an Asian City, Madras,* Social Forces, 1958, Vol. 2, pp. 1-25.
12. C. Rajagopalan, Bombay : A study in urban demography and ecology, *Sociological Bulletin,* 1960, Vol. 9, pp. 19-38.

13. A. B. Chatterjee, Ecological structure of Calcutta twin, *National Geographical Journal of India*, 1965, Vol. 11. pp. 59-62.
14. S.M. Alam, *Hyderabad-Secunderabad (Twin cities): A Study in Urban Geography*, Bombay: Allied Publishers Private Ltd., 1965.
15. U. Singh, Changes in the Builtupur area of Kaval towns in the Ganga Plain (India), *N.G.J.I.*, 1966, Vol. 12, pp. 103-17.
16. R.E. Dickinson, *City Region and Regionalism*, Routledge and Kegan Paul, London, 1964, p. 135.
17. C. Abrams, urban land problems and policies: Housing and town and country planning, *Bulletin* No. 7, United Nations, 1953, p. 34.
18. F.S. Chapin, *Urban Land Use Planning*, Harper and Brothers, New York, 1957, XIV.
19. K.K. Dube, Urban land uses: A critical review, *Indian Geographical Journal*, 1969, Vol. 46, pp. 26-34.
20. G. Breese, *Urbanization in Newly Developing Countries*, Prentice-Hall of India, New Delhi, 1969, p. 58.
21. A.B. Gallion, and S. Eisner, *The Urban Pattern : City Planning and Design*, Affiliated East-West Press, New Delhi, 1965, p. 191.
22. H. Bartholomew, *Land Use in American Cities*, Hardvard University Press, Cambridge, Mass, 1955, p. 15.
23. B.J.L. Berry, and F.E. Horton, *Geographic Perspective on Urban System*, Englewood Cliff, Prentice-Hall, 1970, pp. 444-49.
24. Ministry of Housing and Local Government, The Density of Residential Areas, London: Her Majesty's Stationery Office, 1952.
25. M.P. Collins, Field work in urban areas, frontiers in geographical teaching (Edts. Chorley and Haggett) Methun & Co. Ltd., London, 1965, pp. 217-20.
26. Conzen, 1966, op. cit, pp. 56-78.
27. J.N. Jackson, op. cit., 1968, p. 113.
28. K.S.R. Gowda, *Urban and Regional Planning*, University of Mysore, Mysore, 1972, pp. 86-87.
29. R.L. Singh, *Bangalore, An Urban Survey*, Tara Publications, Varanasi, 1964.
30. B.N. Sinha, *Sirsi—An Urban Study in Application of research models*, Karnatak University, Dharwar, 1970.
31. S.M. Alam, *Metropolitan Hyderabad and Its Region*, Asia Publishing House, London, 1972.
32. L.N. Ram and B. Thakur, op. cit., 1972-73.
33. J.P. Singh, Pattern of land use in urban areas : A case study of Shillong, *Indian Geographical Studies*, Bulletin No. 6, 1976, pp. 25-32.
34. E.W. Burgess, *This Growth of the City*, in R.E. Park, E.W. Burgess and R.D. McKenzie, The City, University of Chicago Press, Chicago, 1925, pp. 47-62.
35. H. Hoyt, The Structure and Growth of Residential Neighbourhoods in American Cities, Federal Housing Administration, Washington, D.C., 1939.

36. R.D. McKenzie, *The Metropolitan Community*, McGraw Hill, New York, 1933.
37. C.D. Harris and E.L. Ullman, The nature of cities, *The Annals of the American Academy of Political and Social Science*, 1945, Vol. 242, pp. 7-17.
38. C. Colby, Centrifugal and Centripetal Forces in Urban Geography, *Annals of the Association of American Geographers*, 1933, Vol. 23, pp. 1-20.
39. H.J. Nelson, The form and structure of cities: Urban growth patterns, *Journal of Geography*, 1969, Vol. 58, pp. 198-207.
40. R.P. Sinha, Dalmianagar : A study of the industrial complex, *Indian Geographical Studies*, Bulletin No. 2, 1974, pp. 26-43.

5

Techniques of Land Use Survey and Mapping

THE PRESENTATION of land use survey techniques is essential for covering large areas within a short time and with considerable accuracy. An inventory land use survey is, thus, basic to land resource assessment. By mapping every piece of land and measuring its characteristics features i.e., soil types, irrigation facilities, modes of land use, land capability, etc., a clue for future possible uses is obtained.

It is an important tool in the hands of those who are engaged in developing the rural areas. The survey, data collection, compilation and the preparation of land use maps require an intimate knowledge of the area concerned. Hence, the land use survey should be accepted only by those who have the will and capacity to do the work in a right-earnest. This survey should be accomplished in the spirit of personal contribution towards rural reconstruction.

MATERIALS REQUIRED

For land use mapping of a revenue village, the investigator requires the following materials:

(a) Cadastral map (16″=1 mile) of revenue village and layout maps (traced copies) to note the important and necessary points available in the field.

(b) Obviously tracing paper is necessary for preparing layout maps from cadastral sheet.

(c) Red, blue, black pencils and pen along with erasive rubber, james clip, adhesive paper, steel pins, cardboard, etc. are the primary requirements at the time of survey.

(d) Varieties of graph papers, pantograph, ediographs are also necessary for reduction and enlargement of maps and stereoscope etc. for reading the aerial photographs of land use.

(e) In India the most suitable times for land use survey on the months of November, December, January, February and March.

PRELIMINARY WORK

Before proceeding to the field for land use study and preparing maps, it is the duty of an investigator to make himself ready to make route march in the field, besides the collection and compilation of huge data relating to the farm characteristics. The investigator should contact local people and farmers who are quite familiar with the village land. This contact should be made in such a way that they must accompany the investigator and cooperate in the field identification for quick progress of mapping. An information about the fertilizers used, crop rotation, per acre yield rate, varieties of crops grown, use of implements and labour force, plus input-output ratio may be gathered with the help of local people. Such people would henceforth be called "local guides".[1] Their cooperation is most valuable in the field survey.

PROBLEMS OF LAND USE MAPPING

The basic question in land use mapping in India is about the technique that should be employed for solving the problems of land use. First of all, the investigator should do the reconnaissance survey of the village with the help of cadastral sheet, overlay map and local guides. The record of land use for each plot of land by direct observation method on large size maps (16″=1 mile) is generally used in most countries of the world. The determination of land use by this method is only suitable when the plots on the maps are sufficiently larger (3 to 4 centimetres) so that the signs and symbols may be easily written thereon. In northern India, the land use mapping is a bit difficult task due to the problems of the division of land into smaller plots and to the varieties of crops being grown in a year on the same plots. In most cases this change is so frequent that no rotation pattern can be easily ascertained. The cadastral maps of the villages bear the same shape and size since 1895-1912 settlement survey. But now with the change in time, no original shape and size of plots are apparently visible. The field is divided into several tiny pieces among the heirs of the same family or owing to the sale and purchase of the smaller prices of the land. The identification of field with the help of layout maps is, thus, difficult at the first site of an investigator. The frequent change in cropping pattern seems apparently to be a problem in the land use mapping due to the changing local conditions. An investigator should never think in terms of these difficulties because our ultimate aim is to prepare land use maps without considering the problem which arises owing to cropping intensity. One visit is not sufficient for such studies. Only frequent visits may yield some fruitful results. For quick work, the idea of crop-rotation, fertilizers used and input-output may also be ascertained by questionnaire method with the help of local people. To avoid the problem of smaller size plots, the easier solution is to divide the village on outline maps into several characteristic regions as the similar condition is available in rural India. The characteristic regions in one village may also be determined with the help of canals, smaller irrigation channels, rivers, higher embankments, roads, footpaths, tanks, knolls, orchards and similar other features that are mostly permanent features of the land for year to come.

In these characteristic regions, construct a group of smaller plots on layout maps by the process of joining thc plots together of a contiguous area but in no case the maximum size of the field should exceed 5 hectares. Every such plot should form a part of the characteristics regions and in no case the outer the boundary will be drawn arbitrarily but that should always follow the boundary of the plots taken together as one unit. Now each such part can easily be identified on the ground and its use may be determined with the help of local guides. Each use must be carefully represented on the traced over lay map (Fig. 5.1) and on alternative with the help of given symbols and signs (Table 5.1). Here the overlay map of land utilization is prepared for Maghra village of Nalanda district.

FIELD INVESTIGATION TECHNIQUE

(a) The traced over lay map from the cadastral sheet is known as field sheet. All mapping on the field sheet should be done in the field and verified with the help of local guide, who should also accompany the investigator for giving the detailed idea of the cropping pattern in a particular field in different seasons.

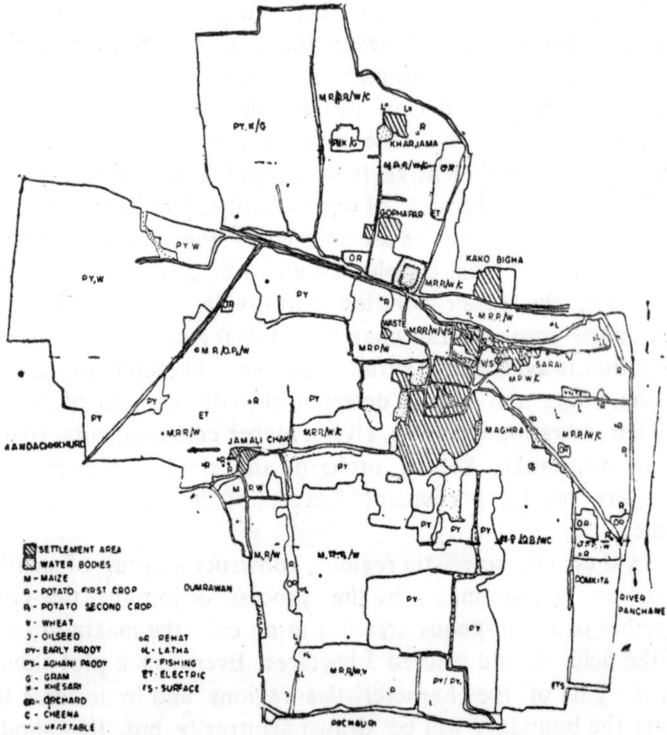

FIG 5.1 : *Maghra : Field sheet for preparation of land use Map 1971 (Nalanda District)*

TABLE 5.1: Symbols and signs for field survey of land use mapping

Land Use	Symbols	Land Use	Symbols
Settlements			
First Paddy	Pd_1	Kutcha Well	
Second Paddy	Pd_2		
Wheat	W	Waterbodies in rainy season	
Gram	G		
Barley (Oat)	B		Blue clour
Pulses	P_1		
Khesari	k	Waterbodies in rainy and winter season	
First Potato	P_1		
Second Potato	P_2		Blue colour
Third Potato	P_3		
Oil Seeds	O	Permanent Waterbodies	Blue colour
First Maize	m_1		
Second Maize	m_2	Sugarcane	Sc
Millets	M	Permanent forest	Pf
Cheena First	C_1	Grassland	G1
Cheena Second	C_2	Summer crops	Su
Ground Nuts	g	Other Bhadai crops	Bt
		Other Rabi crops	Rt
Irrigated land	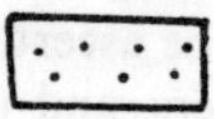	Other Aghani crops	At
		Ancient site & Garh Area	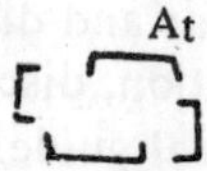
Tube-well	Tb		
		Reserved forest	Re
Pucca-well		Protected forest	(Pr)
		Village Waste land	V1

(b) On the field sheet all the homestead areas must be enclosed with the dotted lines, shaded by the black pencil. The extensions of new housing sites should be clearly marked with black pancil. Ancient sites and abandoned settlements must be recognised from the soil type of (Jhitkaur with high salinity) that particular site. These should be enclosed by symbol (Table 5.1). The present land use of such sites in the cultivable land must be clearly investigated, because quite often such sites have very great archaeological significance in the total continuum of a particular type of civilization.

(c) All public land and lines of communications, e.g., temple and mosque sites, cremation ground, graveyard, road, rail lines, school compound and others should be marked by the red pencil.

(d) All waterbodies, such as river, canal, tank, ditch, pyne, tube-well, drinking well, etc. must be represented with the blue pencil.

(e) The general land use pattern, which includes homestead land, horticulture, orchards, plantation site crop land, culturable waste land, woodland, unproductive and grazing lands etc. should be shown by the black pencil.

(f) After dividing the whole revenue village into characteristic regions of more or less homogenous surface and broad land use classes, the investigator should concentrate on determining the cropping pattern.

(g) An investigator or the party should start the crop survey from one corner of the village, whichever is easier to them. The survey party including the local guide should take one part of the field for investigation at a time. This work must be carefully done and identification should be entered into the map.

(h) The cropping sequence starting from June to the whole year and different seasons must be ascertained by field observation, discussions with the villagers and verification from the local guide. The cropping sequence should be recorded by a combination of letter-symbols in each part of the overlay sheet. If cropping sequence changes from one section of the field to the other, then after dividing the parts into sub-parts on the over-ay map, the cropping sequence should be entered.

(i) Further, the survey party should proceed to the adjoining part and repeat the same process of identification, observation, discussion, verification and the entry of symbols. In case of the same cropping sequence of the two-adjoining parts, the investigator should simply give a sign (Φ) in the middle of the two sub-tracts. Before closing the entire survey one should check that no part of the village is left unchecked.

2. (a) On the margins of the field sheet or overlay map the distribution of the soil type, slope, drainage lines, electric and communication lines, tube-well, drinking well, irrigated area,

area under principal crops, etc. may be represented with small sketch maps.

(b) Some information may also be recorded in a tabular form such as acreage under different crops, yield rate, etc.

(c) Socio-economic data may also be gathered from the villagers after doing household survey of population, caste structure, income group, literacy, etc. so that the investigator may be able to judge the pressure of population on land, socio-economic conditions and the other casual relationship for certain land use pattern.

(d) The information about the future projects if any, along with high yielding crops, etc. may also be noted on the margins of the over-lay map.

(e) North line must be given on the margins of the land use map besides the scale of the map, name of the revenue village, block development office, neighbouring villages in cardinal directions, approach road, nearest market, railway station, district, etc.

VILLAGE SURVEY

Village Survey is an essential part of land utilization survey because it provides the relevant data for analysing the various aspects of the village relating to its spatial distributions and external relations. After studying the varied subjects in geography this field study imparts field survey spirit and research orientation to each and every student. Without intensive field work nothing can be done regarding land use mapping, population distribution and related problems together with any other matters of geographical interest. Such studies are also important in solving the food problems e.g. "Britain used to produce only 35 to 40 per cent of her food requirements in 1939. By 1949 she was producing double the amount". It is the outcome of land utilization survey. In India, on account of financial hurdles, it is essential to complete this great task with the humble beginning of village survey by the graduate and post-graduate students of our universities.

Before proceeding into the details of village survey first of all it is essential to get the cadastral sheet of the revenue village about which the study has to be made. Then reduce that map according to the size of the village survey thesis to show the different aspects of geographical interest. Then prepare a layout map of original cadastral sheet, having location of different tolas, khands, pynes and ahars of the field and other details as given in the following subheads.

Introduction : Location, origin, nomenclature, general shape and size, naming the different parts of the village with its cultural traits, description about Garh site if any, revenue village and tolas, statement of problem, sources of data.

Physiography : General structure of the landscape; higher ground, plain land and low land; slope, water-divide, process of erosion, erosional surfaces, amount of silting and desilting, etc.

Drainage : Surface drainage and underground drainage; surface drainage controls rivers, rivulets, permanent channels, ditch, chaurs, tank, lakes, etc. But only the water level of wells, tube-wells, and surface wells represent the underground drainage except the limestone region and with the help of this the measurement of water level from the surface in summer, winter and rainy season can be made and thus three visits have become essential.

Climate : For climate study, seasonal variations in the direction of wind blow and calm days, rainfall and its intensity and temperature conditions during different seasons; information about storms, Gales and thunder storms are essential. The data about rainfall and temperature may be gathered if the recording station is situated locally, otherwise meteorological directory. for different Development Blocks or Talukas will solve this purpose easily.

Vegetation : Kinds of vegetation (height, density, distribution and use); orchards and groves (e.g. mango, guava, banana, lemon, papaya and bamboo groves) and its relationship with drainage channels.

Soils : Classification of soil-sandy, domat, alluvial, black soil, red soil, reh soil, alkaline soil and jhitkaur, etc. Results of samples of soil test and the study of effects of erosion on different types of soils have been essential.

Land Utilization : See the techniques of land use survey (Table 5.1) land classification and colour scheme.

Irrigation and agricultural water supply : Information

regarding sources of water should be gathered along with the percentage of irrigated area with permanent water channels, canals, tank, ditch, pyne, wells, surface wells, tube-wells, state borings, village etc.

Agriculture : Seasonal distribution of crops and the per cent variation in yield, food crops and cash crops comparison, rotation of crops, crop zones, etc. The information about fragmentation of holding is essential for the cultivators. The conditions of tenancy, manures, seeds, implements, tillage, terracing, bunding, field pattern, condition of livestock, etc.

Village industries : In villages industries like blacksmithy and gold smithy, basket weaving, shoe-making, bidi making, clay pottery, boat making, fishing industry, cotton, tasar and blanket weaving, carpentry, etc. and its origin, locational factors, out-turn disposal and present trend with number of workers and implements, etc. must be considered in any study of village survey.

Trade and commerce : Number and kinds of shops and workers engaged in trade and commerce must be considered under this sub-head. Items of trade, e.g. foodgrains, fodder and essential raw materials for nearby agro-industries from the village and import of consumer goods mainly come under this head. The study of consumer travel pattern from the village to the nearby shopping centre and the total volume of export and import are essential.

Growth of population, castewise distribution of population in different parts of the village, workers and non-workers, age and sex structure, literacy and religio-cultural relationship must be considered for study under this head.

Settlement types, patterns and housing condition : (a) Settlement type of the village should be found out from the dispersed, semi-sprinkled, semi-compact and compact type of village in relation to the intervening space between the houses and housing groups and its effect on bari-land should be studied in detail.

(b) The arrangements of dwelling units and their outer shape should be studied because they may be outcome of conscious planning or natural growth in the shape of rectangular, square, circular, hollow-square, and linear, etc. due to the presence of river banks, temple, church, mosque and orchards, etc. in the centre of the village.

(c) **Type of dwelling units :** Regarding house types it is essential to investigate the materials of wall, materials of roof, single family housing system or joint family housing system, inclination of roof and climatic effects on it. The classification of houses according to number of rooms, number of stories, thickness of wall, nature of ventilation, sanitation, condition of accommodation and comfort should be studied. The ground plan of typical houses, their photographs, siting, grainery store, skyline and human activities related to each type of housing, etc. study is most essential because in this respect each and every village is the emblem of its own.

External relations : It is essential to show the relationship with the nearby shopping centre, thana H.Q., subdivisional H.Q., district H.Q., and Provincial Capital H.Q. A short note about persons employed outside the village and its relationship with the village should also be included. It is also necessary to show the schools, rural health centre, panchayat hall, library facility, milan-mandir, playground, nearby melas, hats and their speciality together with village volunteer force, fire fighting scheme and other facilities are necessary items of study.

Village planning : Problems of rural inhabitants and village development—requirements of the people—planning and zoning for the future—resource management—so that all amenities should be available to the villagers.

TECHNIQUES OF SAMPLING IN LAND USE

In view of increasing demand and various use of the land, it is not possible to survey each and every piece of land over vast territory with few trained personnel and the extent of time and expenditure involved and hence sample techniques seem to be the best for obtainnig factual information in regard to the existing pattern and problems of land use. This would not lead us to a total land use picture relating to actual distribution. It reveals that the land use problems may be similar in areas having homogeneity in regard to physical and social environmental conditions. So far as the beginning of sample survey is concerned in 1886 poverty survey was done by Charles Booth, but sampling was used in 1912 by Bowleg. Similarly in 1924 employment survey was carried on and in 1937-38 household expenditure survey was done.

MAPPING OF SAMPLING

Sample or sampling is a part of the whole more or less representative field of observation. But in a particular study, opinion may differ as to the type of sampling that should be adopted for land use survey. Thus it should be drawn cautiously in the field or a field layout maps, because a faulty sample design leads to a fallacious conclusion.

There is nothing theoretical about the idea of sampling. We live in a world of samples i.e. sample of land, crops, measuring yield rate, intensity of cropping, industrial and institutonal locations, etc. In this way, it is a way or choosing anything from any table, map, population, field of industries as the common asʼects of life. In statistics such picking and choosing requires a scientific and systematic knowledge, so that our observation may not be biased. It must have some logical proof as to which sampling is suited to the study of the particular type of distribution. Thus, sampling has two properties (a) representativeness and (b) small and adequate selection of items.

NEED FOR SAMPLING

Investigator takes the help of sampling as it reduces the unmanageable heterogenous items of observation to a handy one, so that all the types are adequately represented in it. Because of this a sample is defined as selecting a few representative item from the aggregate or total field of study. The results of sampling are reliable as it is mathematically proved that sample mean is the best estimate of the total population mean. It is also useful for measures of location and dispersion.

Advantages of sampling survey over census survey

(i) saving of time,
(ii) saving of money,
(iii) detailed study,

(iv) Accuracy in result,
(v) Administrative convenience,
(vi) Impossibility of the use of Census Survey.

SOURCES OF ERROR

The sample has many advantages over a complete survey or complete enumeration of a finite items of observation. If carefully designed, the sample is not only considerable cheaper but may give results which are just as accurate as, and sometimes more accurate than those of a complete land use survey. The reason is that in survey different ways and efficiency of surveyor is subject to partial error, sometimes considerably larger, owing to faulty sampling method employed. The smaller of the sample makes possible the use of more care in the design and execution of each step in the enquiry. In this way, the sources of error can be investigated and reduced, eliminated, or measured and corrected for the accurate result. Hence, a carefully designed sample may actually be better than a poorly planned and executed land use survey. For highly variable characteristics of results, a sample may have to be so large in order to give the desired results that it is better to use a complete survey or census instead of a sample study. Finally in same problems, such as in crops yield rate measurement, and total requirement of input of fertilizers, only a sample can be used. The sample is only a part of the whole observation. Hence the exact agreement is not possible between sample estimates and actual proportion of land in various uses. The use of any sample in estimates must ensure two properties i.e. accuary and precision.

Accuracy relates to correctness in estimation and if there is consistently over and under estimation of the value then the sample is said to be leased. Any sample procedure should produce estimates that differ from true value only by reason of random disturbance factor.

Precision means the spread of sample estimates of the land use value around the true value. An efficient sample has a relatively small spread of such errors of estimation. Accuracy is the first requirement of any sampling. Accurate procedures and relative efficiency of estimation errors provide a ground for selection or rejection of one method. Only in case of change through time, the biased procedure gives greater efficiency, as in the temporal comparisons the bias may be eliminated.

Disadvantage of sampling survey

(i) chances of biasness,
(ii) difficulties in getting a representative sample,
(iii) need for specialized knowledge,
(iv) impossibility of sampling.

TYPES OF SAMPLING

To fulfil the requirement of investigators, various types of sampling design have been given here, with a view to studying different aspects of land use survey:

Random sample : The haphazard selection of items from continuous distribution is known as random sampling (Fig. 5.2). In this case each and every item has an equal chance of selection, hence it is also known as probability sampling. Samples are drawn with the help of chit or card methods or with the help of random number tables. (Appendix I). In chit method, the required number of chits are taken out and this work may be done 'with placement' or 'without placement' methods.

Stratified random sample : In stratified random sample the land is divided into a number of strata, than a random sample is made for required items within each stratum and thus the different categories or strata of regional land use pattern are represented with the sample. For the application of this method in land use survey the study area should be divided into physiographic units, e.g. uplands, rolling ground, plains, low lands, etc. In this way stratified sampling, first of all requires the stratification of regional land use and only then the selection can be made, so that the results differ as widely as possible from each other. Thus the successful execution of this method is made possible only by using stratification factors and object of the survey (Fig. 5.2).

Systematic sampling : The selection of points at equal interval (Fig. 5.2) for distribution of any item is known as systematic sampling. This method is quite often used by geographers and soil scientists for reconnaissance or grid surveys. In this method before making land use survey, the land has been divided into small grids by considering distance of two places

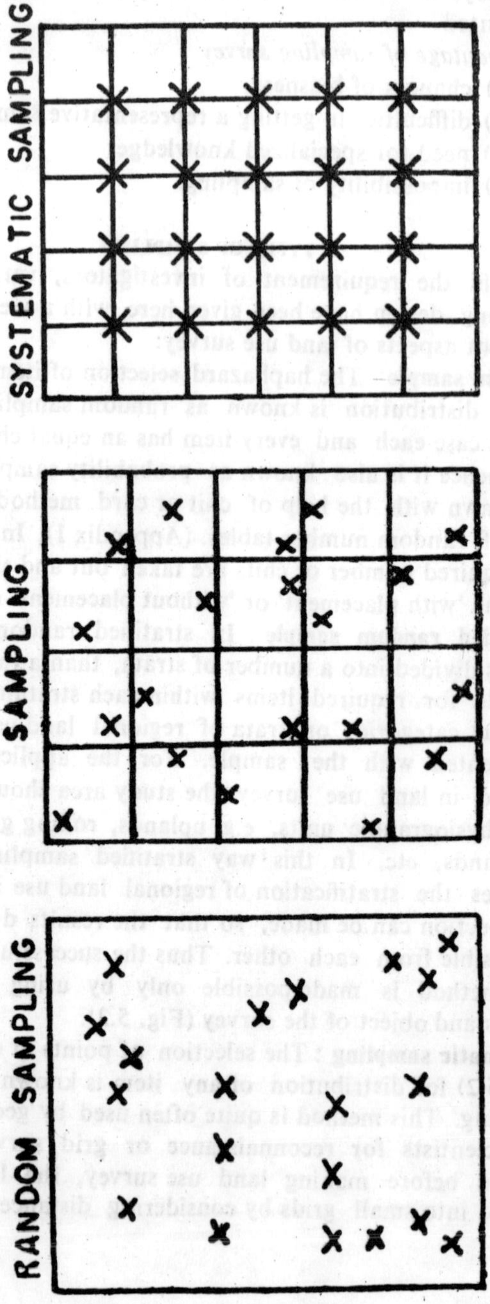

FIG. 5.2 : *Patterns of sampling design*

or latitudes and longitudes over map. It is then superimposed over the land use map of the area and the points are selected accordingly. The grids are not always rectangular and their spacing is, also not arithmetic. Fig. 5.3 illustrates a radial logarithmic grid for sampling the surface slope or public size of an alluvial fan, where distance from the fan head is important, and pebble characteristics are believed to vary logarithmically with distance.

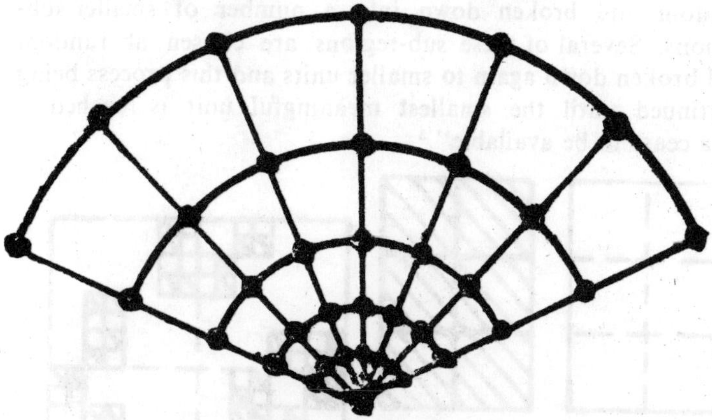

FIG. 5.3 : *Radial logarithmic sampling*

Purposive sampling : In purposive sampling, the samples are taken on the basis so that the representation of each class is made. Once the general break down of the sample is decided on systematic lines, the choice of actual sample unit is left to the investigators, Purposive sampling is, therefore, a method of stratified sampling in which the selection within the strata is non-random. Theoretically, it may bc less generalised than random sampling but with adequate safeguards it can be made highly reliable.

Cluster sampling : Cluster sampling involves first the random selection from groups of observation and each selection leads to cluster of field units and within each of the selected clusters, one can include all the individual units which it comprises or only a sample of them. If a complete enumeration of randomly selected clusters from macro-level to micro-level will be completed, it is known as cluster sampling.

Nested sampling : This sample is useful in exploratory studies. It requires a large region with sufficient local variations, so that grids of major and minor divisions could be easily made (Fig. 5.4). It is also known as multi-level of hierarchical sampling. "Several of these major regions are then chosen at random and broken down into a number of smaller sub-regions. Several of these sub-regions are chosen at random and broken down again to smaller units and this process being continued until the smallest meaningful unit is reached or data cease to be available".[2]

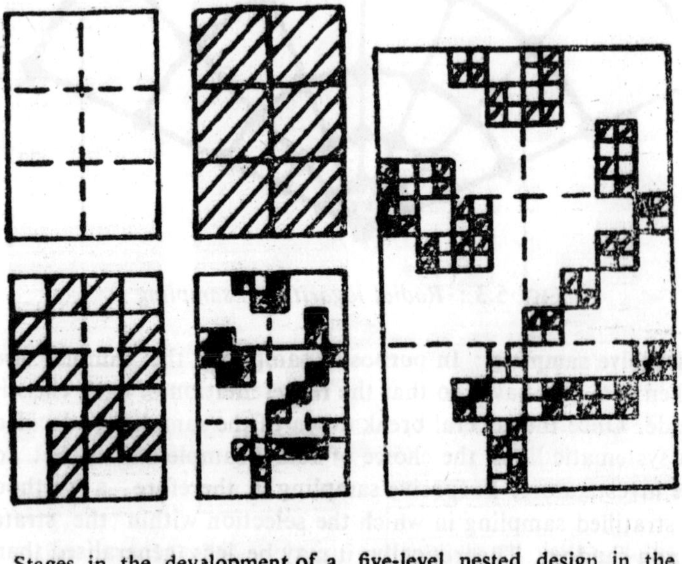

Stages in the development of a five-level nested design in the Tagus-Sado Basin, Central Portugal (G=4.5).
Source: Haggett, 1961, p. 7.

FIG. 5.4 : *Stages of nested sampling design*

APPLIED LAND USE STATISTICS

The use of quantitative approach in most of the standard papers necessitates the review of the statistical formulas which are useful for students and research scholars in bringing the land use studies up to its mark in comparison with the other branches of geography and in noting the advancement of the subject in relation to statistics as a base of all science subjects.

Measures of land potentialities : In India the food production is not sufficient to meet the requirements of growing population. It is a sorry state of affairs that the rural inhabitants suffer from a deteriorating standard of nutrition and that they are undernourished due to the occurrence of flood and famine at intervals. Thus the formula derived by the various geographers[3] are briefly considered here in order to measure the overall potentiality of the land.

1. Agricultural density (Ferenczi, 1938) $= \dfrac{\text{Total agricultural population}}{\text{Total cultivated area}}$

2. Nutritional density $= \dfrac{\text{Total rural population}}{\text{Total cropped area}}$

3. Economic density (Singh, J., 1967) $= \dfrac{\text{Total rural population}}{\text{Total net sown area}}$

4. Intensity of population pressure $= \dfrac{\text{Caloric density}}{\text{Optimum carrying capacity}} \times 100$

5. Caloric density $= \dfrac{\text{Total rural population}}{\text{Total area under food crops.}}$

6. Man-Soil density (kuperus, 1938) or physiological density $= \dfrac{\text{Total population}}{\text{Total cultivated area}}$

7. Weighted average caloric production available for ingestion $= \dfrac{\text{Total caloric output Available for Ingestion}}{\text{Total Area under Food Crops.}}$

Index of crop diversification : In a study of the regional pattern of agricultural land use in India, Bhatia has evolved a formula for determining crop diversification.[4]

$$\text{Index of crop diversification} = \frac{\text{Percent of cultivated Area under X crop}}{\text{Number of X crops.}}$$

Where X crops are those crops which individually occupy ten per cent or more of cultivated area in a district. The lower the value of Index, the higher is degree of cr op diversification. The index value, however, cannot be less than ten.

A Method for measuring and comparing agricultural land use : It has been developed by P.M. Enggass and tested the method in Las Torres of Seville Province.[5]

TABLE 5.2 : **Determination of the index of cultivated area**

Crops	Sown Area in hectares
Green Beans	84.00
Early Potato	84.00
Sugar Beets	67.20
Wheat	78.40
Barley	67.20
Alfalfa	5.60
Olive	9.30
Hurto	*44.80*
Sown Area First Crop	440.50
Maize.	89.57
Sweet Patato	44.80
Late Potato	15.82
Green Alfalfa	5.60
Hurto	*44.80*
Sown Area Second Crop	200.27
Total Sown Both crops	640.77=A
Total Area of Project.	512.13
Less than productive land	41.61
Total cultivable Area	470.52=B

The reforms project of the 'Institute National Decolonization factor=1.34 . . .(unknown)

$$\text{Index} = \frac{A \times 100}{B \times 1.34} = \frac{640.77 \times 100}{470.52 \times 1.34}$$

$$= \frac{64077}{630.497} = 101.63$$

$$\text{Index of yield (Iy)} = \frac{C \times 100}{D}$$

$$= \frac{1080}{9} = 120$$

TABLE 5.3 : **Determination of the index of yield**

Crops	Production (00's kg/ha) A Total obtained	B Previous Project goal	Theoretically sown area	Actually sown
Wheat	25	20	1.25	1
Barley	20	19	1.05	1
Maize	58	24	2.42	2
Beans	10	15	0.67	1
Early Potato	150	117	1.28	1
Clover	400	333	1.20	1
Cotton	18	14	1.24	1
Sugar Beets	380	232	1.64	1
Total	1061	774	10.80	9

Measurement of crop concentration : In order to determine the regional concentration of crops in Uttar Pradesh, an index, the location quotient has been calculated:[6]

$$\text{Index for determining concentration of crops} = \frac{\text{Area of crop } X \text{ in an areal unit}}{\text{Area of all crops in the areal unit}} \div \frac{\text{Area of crop } X \text{ in the entire region}}{\text{Area of all crops in the entire region}}$$

Public agriculture and the chemistry of food production : Public agriculture, as developed by S.S. Bose is a new discipline in Human Geography which highlights certain possibilities and impossibilities in the matter of food production. The circulation of each of the thirteen fertilizer elements, *N*, *P*, *K*, *Ca*, *Mg*, *S* . . . in nature is governed by the equation:[7]

$$(K_1-K_2)\ P=G+A$$

Where $K_1=K_2$ is a constant depending on the food habits, conservancy practices etc. of the community.

P= the population.

G= Soils annual quota of fertilizer elements from natural sources, and A is the quantity of artificial fertilizers.

If $G+A$ falls short of $(K_1-K_2)\ P$, the use of high yielding varieties cannot, in view of the operation of the law of conservation of Mass-Element-wise, Lieleig's law of the minimum and the law of Diminishing Returns, helps in producing the food required by the country.

Index of land consolidation : The holding management is only advantageous when the farmstead is in the centre of a circular holding. This is expressed by the formula:

$$S=\frac{r}{2}$$

Where S is index of land consolidation, r is radius from the centre of a circular holding.

The square shape of farm is equally advantageous. There are several formulas for computing the optimum shape of farm. One of them is Kosicki's formula.

$$U=\frac{Lr}{Li}$$

Where U is the index of the configuration of the holding, Lr is the actual average distance between farm land and farm stead, Li the ideal distance calculated on a holding of the same area in the shape of a circle or square.

Moszczanski's formula:

$$U=\frac{O.S}{P}$$

U is the index of the configuration of the holding. O is the circumference of the farm, S is average distance of the farmstead, P is the farm area. The average distance is calculated with the assistance of a square network.

The above formula indices may be supplemented with one more very essential index[8] depicting the degree of land consolidation.

$$K=\frac{\dfrac{\sum_{i=1}^{n} a_i}{n}}{\sum_{i=1} a_i}=\frac{a_1+a_2+a_3\ldots+a_n}{a_1+a_2+a_3\ldots+a_n}$$

Where a_1, a_2, $a_3 \ldots a_n$ are symbols of various plots of different size.

If the index value is l, the holding is composed of a single plot. Indices of holding composed of several plots are included between O and 1. In practice indices are multiplied by 100 to obtain percentages of consolidation. It is useful in showing changes which have occurred in a given period of time. It is a relative value independent of the size of the holding. The index of consolidation may be supplemented with the dispersion of plots and it is calculated with the help of following formula:

$$R=100-100K=100l(l-K)$$

Distribution of dividends over farm land : A formula has been devised for justifiable distribution of dividends on the basis of participants contribution to the farm in the forms of land, labour or capital. The application of this formula will show that the income of each family or cultivator within the farm will be much more than his present income from the individualistic operation of land holding. The formula is as follows:[9]

$$D=\frac{T-E}{M\times 100}(L+I+C)$$

Where D=Dividend to be obtained by each participant.
T=Total earning of the farm.
E=Total expenditure of the farm on various heads.
M=Total number of member participants in the farm.
L=P.C. dividend of each participant for his land.
l=P.C. dividend of each participant for his labour.
C=P.C. dividend of each participant for his capital share in the farm.

If each participant had possessed equal amount land in the farm, offered labour of equal value and contributed equal share of capital, then it would have been easier to find out the dividend per head to be distributed equally. There will be variance in respect of the shares of each participant land, labour and capital. For the fact, the above formula can be modified in the following way for the maintenance of justice from all sides in the distribution of dividend.

$$D=\frac{T-E}{M\times 100}\left(\frac{L\times LS}{Lavg}+\frac{I\times Is}{Iavg}+\frac{C\times Cs}{Cavg}\right)$$

$$D=\left(\frac{L\times Ls+I\times Is+C\times Cs}{Ls+Is+Cs}\right)\frac{T-E}{M\times 100}$$

Where *Ls* = Amount of land of a participant member.

Lavg = Average amount of land of the participants in the farm.

Is = Amount of labour offered by a participant in terms of time.

Iavg = Average amount of labour offered by the participants in terms of time.

Cs = Amount of capital share of a participant.

Cavg = Average amount of capital share of the participants.

According to M.N. Das, "In Galia village of Kamrup district in Assam the total land is 850 bighas. In case of a joint farming the total income from this farm would have been about Rs. 5 lakhs. If Rs. 1 lakh is spent in various activities of the farm including the activities of social welfare. of the villages, then Rs. 4 lakhs will remain to be distributed among the participants of the farm. If all the participants offer equal amount of labour in all the working days of the year if the average land per head in the farm is 17 bighas and if the average amount of capital share per head is Rs. 800, then the annual dividend of the above farmer may be calculated as follows:

$$D=\text{Rs.}\frac{500{,}000-100{,}000}{50\times 100}$$

$$=\left(\frac{30\times 15}{17}+\frac{50\times 365}{365}\times\frac{20\times 500}{800}\right)$$

$$=\text{Rs. } 7{,}117.00$$

Therefore, the above farmer will get annual income of Rs. (7,117.00−2,000.00)=Rs. 5000.00 more than what he would get from the same amount of land, labour and capital, if he would operate his land holding individually."

Changing patterns of crop land use : The patterns of crop land use in each of the sample years are studied in two ways:[10] by ranking of crops according to their intensity and by deriving crop associations on the basis of Weaver's formula.

$\left(\frac{\Sigma d^2}{n}\right)$ as applied in its modified

form by Doi (Σd^2). A comparison of these two aspects in different years gives a true assessment of the evolution of land use patterns in a region.

Agricultural dynamicity : (1) The calculation of Index values of total volume of change by the following equation:[11]

$$IC = \frac{V}{An} \times 100$$

Where IC is the Index of total volume of change, V is the value of total volume of change and An is the proportion of acreage under crops to gross cropped area.

(2) The compilation of index of dynamicity on the basis of the average index values of total volume of change for each pargana is calculated by the following equation.

$$ID = \frac{AVU}{AVR} \times 100$$

Where ID is the index of dynamicity.

AVU is the average index value of the unit and

AVR is the average index value of the Region.

Location quotient for agricultural employment : "The location quotient measures the degree to which a specific region contains more or less than its share of employment in any particular industry briefly indicating a ratio of ratios. To calculate the ratio of employment in agriculture to total populations the formula is:[12]

$$\frac{\text{Employment in agriculture}}{\text{Total Population}} = \frac{7,382,009}{20,155,600} = 36.6\%$$

The ratios of these two variables of each district have been calculated and then by dividing these district ratios with the State percentage of 36.6 the location quotient for each district was found out".

Agricultural deficiency regions : To measure the agricultural deficiency regions of India, output per unit area has been taken as the basic tools for consideration. In this particular case per hectare yield of twenty five crops of the year 1963-64 has been taken into account. All the twenty five crops have been grouped into cereals, pulses, oilseeds and money crops. While per hectare yields of pulses, cereals and oilseeds have been computed in tonnes, for money crops, all the yields have been converted into values (Rs.) as per the ruling price for that particular year.

Mean figures for India's whole have been found out for cereals, pulses, oilseeds and money crops. Thus standard deviation formula has been applied for cereals, pulses, oil seeds and money crops. After that the standard score for each district of India has been computed by adopting the well known formula.[13]

Standard Value $Zic=\frac{ci-c}{\sigma c}$ for cereals

$$Zip=\frac{Pi-P}{\sigma P} \text{ for pulses}$$

$$Zio=\frac{Oi-O}{\sigma O} \text{ for oil seeds.}$$

$$Zim=\frac{Mi-M}{\sigma M} \text{ for money crops,}$$

where for example σC is equal to $\sqrt{\frac{\Sigma(Ci-c)^2}{n}}$

The standard for *Zic*, *Zip*, *Zio* and *Zim* are calculated. To measure the volume of production *Zi* values have been multiplied by the acreage figure i.e. $Zic \times AC$, $Zip \times AP$, $Zio \times AO$ and $Zim \times AM$.

These values are arrayed in descending order for plus values and ascending order for minus values separately. Quartiles are found for all the figures with plus and minus values which are above and below the country's average respectively. For this paper below (–) values have been taken into account.

At the end to determine the total aggregate efficiency of the component areal unit the following formula has been adopted in case of all the districts to find out the weighted average of *Z*'s.

$$\bar{Z}=\frac{(Zic \times AC)+(Zip \times AP)+(Zio \times AO)+(Zim \times AM)}{AC+AP+AO+AM}$$

All the $\bar{Z}$ minus values indicate the below the country's average while $\pm$ are equal to the average efficiency of India.

All the values are mapped and deficiency regions so far as cereals, pulses, oilseeds and money crops and also aggregate deficiency are mapped.

Efficiency measurement : The following statistic is suggested as a way of making a comparative evaluation of the efficiency with which various countries utilize their total food resource environments.[14]

$$\frac{A^1}{A}(y_1,n_1, a_1+y_2n_2a_2+ \ldots .y_nh_na_n),$$

Where: A is the total cultivated area.
A' is the adjusted food producing area.
Y is yield relative for each food.
n is the nutrition relative for each food and
a is the area relative for each food.

The ratio A^1/A indicates the change in a country's cultivated area through the use of noncultivated land, nonland food resources, food imports or exports, etc. The quantity inside the parentheses combines the diverse components that affect food availability.

The computations indicate that Japan uses its food resource environment ten times more efficiently that the United States. The statistic also may be adopted for use as a measure of population pressure and economizing inputs into agriculture for food production.

Index of disease concentration : For each disease a concentration index can be calculated for any areal unit and the formula is as follows:[15]

$$C= \frac{d}{S} \times 100$$

Where C=index of concentration for disease X in a district.
d=percentage of persons suffering from disease X in a district to the total number of disease sufferes from the district.
and S=percentage of persons suffering from disease X in all the districts to the total number of persons suffering from all diseases of the State.

If the index value derived from the above formula is

greater than 100 the district has a share greater than it would have had if the distribution were uniform in the entire State and therefore, it points to the fact that the district has a concentration of the particular disease. All the index values that show concentration are then arranged in a descending order and divided into five equal parts to show very high, high, medium, low and very low degree of concentration.

Localization of crops : The locational coefficient shows the measure of concentration of any entity and has been used by industrial economists to demarcate areas of industrial concentrations and has also been used to indicate localisation of crops. Formula is as:[16]

$$LC=\frac{A}{P}$$

Where *P* is the percentage of a particular crop of the district in the total crop production of State; and *A* is the percentage of area of the district to the total area of the State. If the locational coefficient for a district is (1) one means that the district is having just the production of a particular crop which it should have.

Crop intensity : The procedure adopted for arriving at the intensity of cropping per year is discussed below:

Let *Aijk* and *Cijk* be respectively the cropped area and the area left fallow in the year, *jth* holding and *kth* season denoted by *Pijk* was obtained as:[17]

$$Pijk=\frac{Aijk}{Aijk+Cijk} \quad \ldots (1)$$

and crop intensity in *ith* year and *jth* holding by

$$Pij=\sum_{K=1}^{2} \frac{Aijk}{Aijk+Cijk} \quad \ldots (2)$$

Intensity of cropping in *jth* holding for each period was worked out by averaging out the corresponding *Pij*'s over the years to which the period pertained. This method takes area under the perennial crops as double cropped. The numeration of expression (2) equals gross cropped area in the *ith* year, *jth* holding and the denomination equals net area sown plus cultivable area left fallow in both the crop seasons. Applying this method, intensity of cropping in different holding size classes was obtained.

Mechanization in the employment of human labour : The following models were used to study the various stages of mechanization in the employment of human labour:

$Lij=Di+eij$ . . . (1)

$Lij=Di+bxij+eij$ . . . (2)

Where *Lij* represents the employment of human labour (family/casual/hired/total) per hectare on the *jth* sample form of the *ith* farm category and *xij* is the farm size. And *Di* is the mean.

$i=O$ for bullock farms.
$=1$ for bullock+tubewell/oil engine
$=2$ for bullock+tubewell and other farm machinery.
$=3$ for tractor+tubewell and other farm machinery.

Both the models were used. However, model (2) which eliminated the linear effects of farm size did not show any improvement in the results, as the correlation coefficient between farm size and the employment human labour per hectare turned out to be non-significant. This indicates that by transforming the variables to per hectare basis, the farm size effect is reduced to a large extent. Therefore, the results presented in the discussion refer to analysis of variance only.

In the above models, the differences $Di-Do$ ($i=1, 2, 3$) and $Di-Dk$ ($ijk, t, k=0, 1, 2, 3$) represent the effect of mechanization stages on the employment. These differences were also subjected to the test of significance by calculating the critical differences.

Percentage change in the employment of human labour was worked out as $\frac{Di-Do}{Do}\times 100$. Share of the *jth* farm operations to the total change in employment was calculated as:[18]

$$\frac{Dji-Djo}{Di-Do}\times 100$$

Where $i=1, 2, 3$ (represent per hectare).
$j=1, 2 \ldots$ (farm operations).

Impact of hydrology on land utilization : An elaborate discussion about the correlation of different hydrological character and land use items of the sub-divisions has been had after an study of the precipitation, drainage, floods and droughts, river sub-basins and ground water properties and extraction in the region. Each sub-division has a pattern=difference or *a* difference in intensity of monopattern form which has been determined by *a* deviation from a standard set for the region. The standard is obtained by applying the formula:[19]

$$S=\frac{(a_1+a_2+a_3+\ .\ .\ .)\times 100}{t}$$

Where a represents the standard a_1, a_2, a_3 . . . areas under different items at "t" the total area of the region.

Measurement of spatial discontinuity : An index in land tenure : The basic index outlined here may be referred to as the relative index of land. Percentization, *Pi*, and is given by the equation.[20]

$$Pi=\frac{\frac{1}{S}\ Dt}{100}$$

Here *Dt* is the aggregate distance between land fragments: the actual total distance covered by the operator in a single round that takes him to all his parcels, and *S* represents the hectarage of each parcel. Instead of total land area, average parcel size is used for each land fragment is in effect a separate firm or establishment especially where large distances are involved (where there is no fragmentation $Pi=0$).

Index of water requirements to cultivated area : Potential evapotranspiration as a basis can be used for reckoning the need of water in cubic meter per square kilometre of the cultivated area, after taking into account the actual rainfall. It has been done as:[21]

Water requirements per sq. kilometre of the cultivated area.

$$\frac{\text{Potential Evapotranspiration}-\text{rainfall}}{100}\times 10^6 \text{ cubic metres.}$$

Where 100 is used for the conversion of cms. into metres, 10^6 to change square metres to square kilometres.

Nutrient indices : This has been calculated for different soil nutrients i.e. phosphorous, potash organic carbon based on the percentage of the soil samples tested, which fall in the low, medium and high categories for a particular nutrient. The calculation is based on the following formula:

$$\frac{\text{Low}+2\times\text{Medium}+3\times\text{High}}{100}$$

The obtained figure is compared with the following three given indices for low, medium and high which will coincide in any one of the indices.[22]

LAND POTENTIAL MODELS

Most of the models have been developed for areas of shifting cultivation of Africa. Allan (1965) gives an exhaustive theoretical survey of these areas and defines the land use factor which forms the basis of many of these models. This factor is defined (Allan, p. 30) as the relationship between the duration of cultivation on each of the land or soil units used in classification and the period of subsequent rest required for the restoration of fertility. The actual acreage of land cultivated per head of population he defines as the cultivation factor.

Gourou (1962) suggests that the average density of population in Black Africa is only 25 per square mile of total area—but 750 per square mile of cultivated area. In order to define the potential density permitted by the various types of shifting agriculture observed in Africa he suggests the following formula:

$A . C/B$

Where *A* equals the amount of cultivable land expressed as a proportion of the whole, *B* equals the total length of rotation in years (period of cultivation plus period of fallow) and *C* equals the number of inhabitants per acre cleared annually. If we take *A* at 0.8 (80 per cent), *B* at 8 (say one year of harvest and 7 fallow or 2 years harvest and 6 year fallow) and *C* at 4 inhabitants per acre cleared each year then

$A.C/B = 0.8 \times 9.8 = 0.9$ (overall pottential density per cent measured in inhabitants per acre, or 256 to the square mile).

In such a formula were applied to the whole of Black Africa it would appear that population density could increase ten times from 25 persons per square mile to 250. Obviously the application of such a formula demands precise knowledge of the existing relationships between population and cultivable land. Regional variations in soil fertility and cultivation techniques would play an important role in determining potential patterns.

Another Belgian geographer Beguin (1964) develops several more complicated formulae for assessing land carrying potential in the Kasai province of the Congo. He defines his potential production as v, the maximum agricultural production from a given area normally obtainable under the existing agricultural system. This potential varies with the agricultural system and the physical environment. If c is the years of cultivation and j is the years of fallow then $c+j$ equals the years of the agricultural cycle. Each year part of an area changes from cultivation to fallow so a given population needs $c+j$ units of area. Each year also c units of production are harvested and the annual agricultural production equals P. The production from one unit of cultivated area is

$$\frac{P}{c+j} \text{ and therefore } v=\frac{P}{c+j}$$

Beguin states that in Kasai yields per hectare are 0.5 tons for groundnuts, 0.3 tons for beans and 3.7 tons for cassava. Therefore in a rotation of three crops

$$P=0.5+0.3+3.7=4.5 \text{ tons}$$

Land is cultivated for $2\frac{1}{2}$ years and lies fallow for $6\frac{1}{2}$ years.

$$\text{Thus } c+j=9 \text{ and } v=\frac{4.5}{9}=0.5 \text{ tons per hectare.}$$

POTENTIAL FOR AN AGRICULTURAL SYSTEM

In most tropical agricultural systems the number of crops is for greater than in the first model and there are usually several interlocking rotations. So far his second model Beguin includes multiple rotations. Then the model may be stated as follows:

$$V=\frac{\sum_{i=1}^{l} vi.\,ri}{\sum_{i=1}^{l} ri}$$

Where l is the number of rotations vi is the potential of the ith rotation and ri is the area occupied by rotation i. When this formula is applied to data from Kasai where there are six principal rotations. The results are as follows:

Vi=0.5, 0.47, 0.3, 0.3, 0.42, 0.42 tons per hectare.

When weighted by the area assigned to each rotation:

$$V=\frac{(0.5)\,32+(0.47)\,12+(0.3)\,25+(0.3)\,6+(0.42)\,7+(0.42)\,12}{32+12+25+6+7+12}$$

$=0.41$ tons per hectare

or$=41$ tons per sequare kilometre

POPULATION DENSITY POTENTIAL

This varies according to the standard of living of the population and Beguin's third model takes this into account. If u is the unit of production per capita and if the population desire one ton each per year then they would need $\frac{u}{v}$ land units. The area (S) necessary for a given population (P) so that each inhabitant could obtain a certain annual production (U) practising a certain type of agriculture on a given soil (V) would be

$$S=\frac{pu}{v}$$

Then the maximum density (d) of this population would be

$$d=\frac{P}{S}=\frac{V}{u}$$

for example if $v=60$ tons per square kilometre in an area where there are 1,000 people producing one ton each then

$$S=\frac{Pu}{v}=\frac{1{,}000.1}{60}=16.67 \text{ square kilometres}$$

If $S=20\ \text{km}^2$ then $P=\frac{20\times 60}{1}=1{,}200$ people.

If $u=1.2$ tons then $d=\frac{60}{1.2}=50$ inhabitants per square km.

AREAL DIFFERENTIATION

It is more realistic to assume that the potential varies through area in response to environmental and cultural differences. If we accept that the maximum (VM) and minimum potential (Vm) are on a continuum and vary linearly then mean potential ($\bar{V}$) equals $\frac{VM+Vm}{2}$.

Then in a region of 100 square kilometres with a potential varying linearly from 100 tons per square kilometre to 60 tons per square kilometre, giving one ton per inhabitant per year the population maximum would be

$$P=\frac{100\times\frac{100+60}{2}}{1}=8{,}000 \text{ inhabitants}$$

Thus the maximum population density is 80 inhabitants per square kilometre.

Beguin develops several more refined formulae[23] which take into account sequent occupation of land in which later occupants are forced on to land of lower potential, non-homogeneous regions; input of work per capita; and change over time as it affects population, production per capita and potential. Thus the model is both dynamic and wide ranging as to its input.

REFERENCES

1. P. Dayal, and A. Sharan, Land use patterns in the Bihar Sharif area, Patna District, Bihar, *Symposium on Land Use in Developing Countries*, 21st International Geographical Congress, 1968, Aligarh, 1972, p. 150.
2. R.J. Chorley and P. Haggett, *Frontiers in Geographical Teaching*, Methuen & Co. Ltd., London, 1970, pp. 175-177.
3. Jasbir Singh, Optimum carrying capacity of land, caloric density and intensity of population pressure changes in Punjab, 1951-61, *The National Geographical Journal of India*, Vol. XVII, Part I, 1971, p. 40.

4. Harpal Singh, Crop diversification in the Malwa of Punjab, *The Indian Geographical Journal*, Vol. XXXVIII, No. 1, 1963, pp. 75-78,
5. P.M. Enggass, A method for measuring and comparing agricultural land use: An Andalusian experiment, *The Professional Geographer*; Vol. XIX, No. 3, 1967, pp. 125-129.
6. Majid Husain, Patterns of crop concentration in U.P., *Geographical Review of India*, Vol. XXXII, No. 3, 1970, p. 170.
7. S.S. Bose, Human geography, public agriculture and the chemistry of food production, *Proc. 55th Ind. Sc. Cong.*, Part III, Abstract, p. 260.
8. Jozef Janceszewski, Index of land consolidation, *Abstract of Papers*, 21st, I.G.C., 1968, pp. 144-45.
9. M.M. Das, Plea for joint agricultural farming: A case study in rural economic geography, *Journal of the North-East India Geographical Society*, Vol. V, No. 1 & 2, 1973, pp. 47-56.
10. M.F. Siddiqui, Changing patterns of crop land use in Bundelkhand, India, 21st I.G.C., 1968, Abstract of Papers, p. 181.
11. R.R. Tripathi, Agricultural dynamicity and efficiency levels: A Case study of the Upper Ganga-Gomati Doab, *The Indian Geographical Journal*, Vol. XLVII, No. 1 & 2, 1972, pp. 12-19.
12. N.L. Gupta, S.L. Hiran, Agricultural regions of Rajasthan, *The Deccan Geographer*, Vol. XI. Nos. 1 & 2 1973, p. 61.
13. B.N. Sinha Agriculture deficiency regions in India, *Proc. 57th Ind. Sci. Congress*, Part III, Abstract, p. 227.
14. L. Zoblar, A proposed food resource efficiency statement for the United States and Japan, Abstract: *Annals of the Association of American Geographers*, Vol. 49. 1959, p. 222.
15. M.F. Siddiqui, Concentration of deficiency diseases in Uttar Pradesh, *The Geographer*, Vol. XVIII, 1971, p. 94.
16. Sundrama Sastry, Location of crops, *The Indian Geographical Journal*, Vol. XVII, 1942, pp. 110-114.
17. P.N. Soni, Cropping pattern and crop intensity in various size class of farmers in some I.A.D.P. Districts, *Agricultural Situation in India*, October, 1974, p. 484.
18. H.S. Bal, *Impact of Mechanization on Farm Labour Employment, Agricultural Situation in India*, Sept. 1974, pp. 385-86.
19. S.L. Ambastha, Hydrology and Its Impact on the Land utilization of Magadh, Bihar, *First Indian Geographical Congress*, New Delhi, 1972; (Bibliography Hydrology).
20. M. Uzolgleozurike, Land tenure relations, social relations and the analysis of spatial discontinuity, *Area*; Vol. 6, No. 2, p. 133.
21. Jasbir Singh, Seasonality and Regionalism in the Agricultural Meteorology of Hissar District, Haryans, *The "Geographical Review of India*, Vol. XXXIII, No. 3, 1971, p. 172.

22. S.K. Sadhukhan, A Quantitative analysis of an agro-irrigation facet of West Bengal, *The Geographic Review of India*, Vol. 36, No. 3, 1974, pp. 196-203.
23. P. Gourou, Agriculture in the Africa: The observations of a geographer," Paper Read at the University of Oxford, 1962.

6

Logic of the Formulation of Models of Land Use

THE REPRESENTATION of the real world situation in simplified form is known as a model. This may be an analogy, an ideal, a description, an abstraction, a generalization, a form of refrence, a programme of research, an explanation of how a system works, a theory, a demonstration in the process of simplification of complex situations in a mathematical or diagramatic form. The process of model building seems to be a descriptive cum nomothetic approach in land use studies which explains the complex perception of distribution of land use pattern in a systematic way. In this chapter the purpose is to analyse the models used in land use studies and their formulation in both spatial and temporal dimensions. The variation in land utilization is generally determined according to the changing needs of the society in space and time. The simplification of the real world situation of land utilization through studies either from category, natural regions and socio-economic models is highly helpful in the formation of the concept of land use. First of all Von Thunen had made generalizations about the land use intensity in rural areas. But recently the general land use model, diagramatic model, models of optima and limits, the polar and anti-polar models of land use and gravity models are considered here in detail.

The category model: A study about the types of land use is known as category model, e.g., five categories of land use are found in India i.e., net sown area, fallow, pasture, forest and waste. In the same manner the First Land Use Survey of Britain conducted by L. D. Stamp records seven typology of land use while Second Land Use Survey conducted by Alice

Coleman recorded seventy categories. Hence the Second survey has lost the inhernt meaning concerned with each other in the map analysis. This expedites that the fewer categories are more helpful for the drawing of maps in land use studies whereas larger categories are helpful in data analysis only.

The natural regions model: The regional patterns of land use are the product of relief, climate and structure of rocks, e.g. on the Malabar coastal plains of India the cultivation of rice and coconut are prominent land utilization of the area where as the northern Konkan coast is covered with dense forest because this is the rugged terrain of Sahyadris which extends up to the sea shore. One may hope that this sort of land use should prevail all over the world in the chraracteristic regions but due to variations of climate and relief features of landscape the same is not possible in every ecological condition. This needs some modification of the observed discrepancies by human ingenuity for the best use of the land.

Socio-economic model: It encompasses the influence of the commune, income of the people, technical abilities, transportation linkages, types and patterns of settlements, power stations, etc., on the utilization of land of an area. But in awayance with the structure, relief and climatic conditions of an area the cultural features may not play a vital role in shaping the types of land use. Hence this model regarded as partial rather than fullfledged.

LAND USE MODEL OF VON THUNEN

In the study of agricultural land use, it has been found that the ecological balance of agricultural land in relation to transport nodes, location of cities and rural settlements in the surrounding areas play significant role in determining the farm production to be raised in different areas. Von Thunen was a German economist who analysed the spatial distribution of farming systems, kinds of produce to be raised at a given place and the influence of transportation lines in the localization of agricultural production. Further he considered a city in the centre of an isotropic plain.[2] In this way Von Thunen proposed his concept over an isotropic plain which have no trade connections with the outside world and was surrounded by the uncultivated wilderness. In case the soil is of uniform fertility, the city should be located centrally surrounded by the rural settlements which have no improved means of transport. On the basis of production of goods, distance travel to the market for particular price, and calculated economic rent on each type of land use from the central city to various distances, he obtained on idealised condition for the distribution of production zones as a series of concentric circles from the city centre. (Fig. 6.1)

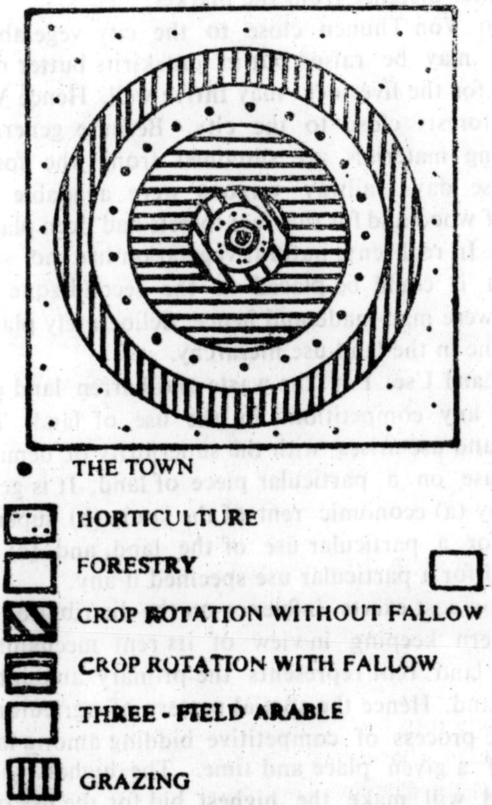

FIG 6.1 : ***Von Thunen's System: Original model of land use types***

He also observed that the land near the market shall have large inputs of manure and labour, etc., with intensive methods of cultivation, while it will be extensive away from the market. Two exceptions to this rule are as follows:

(i) In case per hectare production is large despite small inputs as in forestry,

(2) In case large quantity of inputs yield a small amount as ghee at a smaller distance from the market.

According to Von Thunen close to the city vegetables, berries and milk may be raised but in outskirits butter root crops and grass for the livestock may thrive well. Hence Von Thunen placed forests close to the city. Because generally fuel and building materials are obtained from the forest enbulk. In those days railway wagons were available for transportation of wood and for this fact forests had been placed near the market. In real sense natural vegetation has no such importance that it could be placed in the second zone but here the forests were man-made and hence deliberately placed in the second zone in the land use hierarchy.

Competition of Land Use: For the waste and barren land one cannot observe any competitions in the use of land. The competition of land use arises with the superiority of demand for a particular use on a particular piece of land. It is generally measured by (a) economic rent of the land, (b) superiority of demand for a particular use of the land, and (c) the character of land for a particular use specified if any.

Transportation cost exerts influence on the distribution of of land use pattern keeping in view of its rent mechanism. The amount of land rent represents the primary amount of product on the land. Hence the spatial pattern of agricultural product on is the process of competitive bidding among land use occupancy of a given place and time. The highest yield per unit of land will make the highest bid for the use of a particular plot of land. Such rent referred by Von Thunen was more specific—"Economic rent, which is the surplus of income accruing to a unit of land above the minimum income required to being a unit of new land into use at the margin of production."[3] This differenee is due to spatial variation of two places in terms of fertility and distance from the market in Von Thunen model.

Economic rent can be estimated with the help of the following formula.[4]

R=E (P—a)—Efk

Where R=rent per unit of land,

K=distance to market,

E=output per unit of land,

P=market price per unit of commodity,

a=production cost per unit of commodity (including the farmer's payments to himself for his own labour).

f=transport rate per unit of distance for unit of commodity.[5]

Figure 6.1 illustrates the Von Thunen's ring formation as the hypothetical case in which five land uses compete with each other for space around a settlement.

Later on, change in the original concentric pattern were observed if the initial assumption of an uniform transport system and a single urban centre is relaxed (Fig. 6.2). The

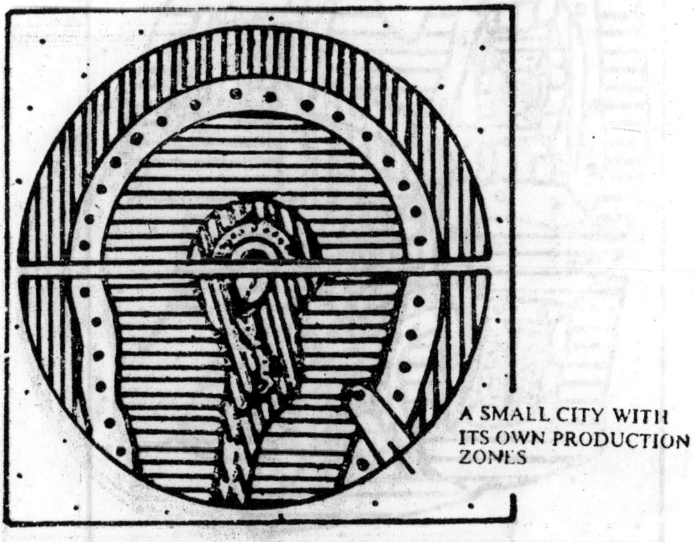

FIG. 6.2 : *Von Thunen's system of land use with navigable waterway and satellite town*

inclusion of a transport line would be beneficial for farmers closest to the same. In this way the original land use patterns would change parallel to the transport routes. The location of a competing market would result in the formation of land use zones around that centre. In riverside location of a city the effect of navigable waterways can easily be seen (Fig. 6·3).

Further, Von Thunen assumed a river flowing across the plain of the city. These new transportation lines extend the concentric zones in both upstream and downstream. In upstream it will be more marked, because downstream movement would be faster and easier (Fig. 6.3). As for example, this type of land use is found along the southern bank of the Ganga near Patna, Buxar, Varanasi and Allahabad in the plains of North India. Now-a-days with the invention of steam engines,

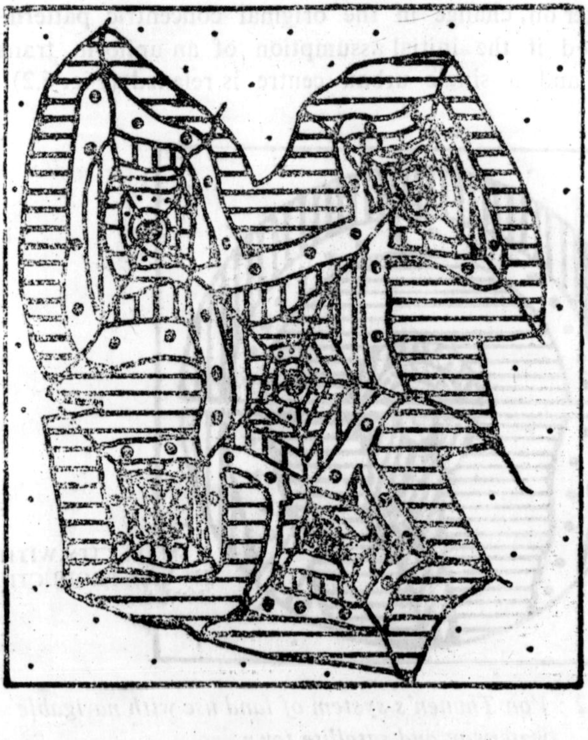

FIG. 6.3 : *Von Thunen's model with improved transport routes and several competing centres*

upstream river transportation is almost as fast as downstream movement. In modern economies, streamlined railroad passenger service, e.g. paved highways, motor truck lines and airways have extended the zones of the city along the lines of transportation arteries. Hence 'modern transportation and refrigeration techniques have overcome perishability and bulk limitations',[16] and for this fact Von Thunen's concentric ring of circles no longer restricts the city areas which provides goods and services to the surrounding zone of influence.

In the same way, if a large number of hamlets had appeared on the plain the resulting pattern of agricultural land use would probably have looked like multiple nodes of intensive land use (Fig. 6.4). Here the hamlets are seem to be of varying population size, reflecting considerable difference in the arrangement of land use. It is directly influenced by transporta-

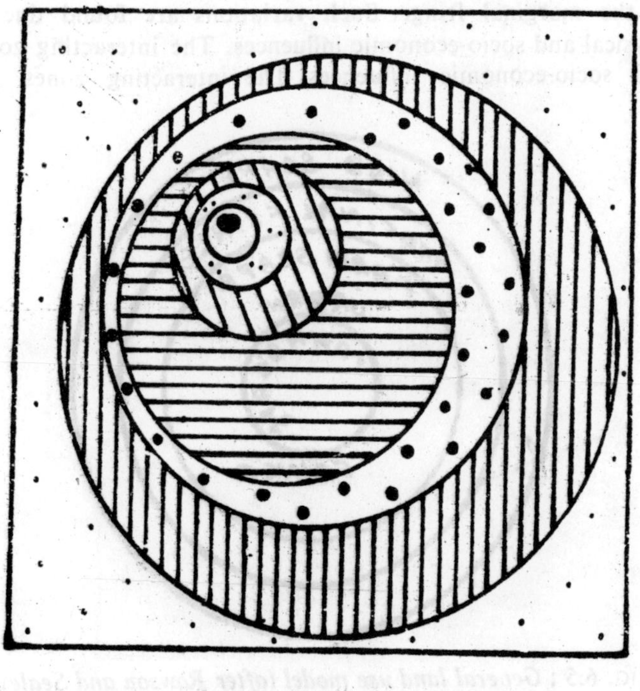

FIG. 6.4 : *Agricultural land use model*

tion arteries. After the relexation of Von Thunen's initial assumptions of uniform production costs, the area at the lower left (Fig. 6.4) has acquired some advantage over area at the upper right in terms of production cost. It makes possible for the farmers in the favourable area to absorb highly transport cost in comparison with the producers of least developed area. In this sense proper use of agricultural land expands at the expense of the other use.

GENERAL LAND USE MODEL OF COLEMAN

Alice Coleman[7] suggested this model in 1969 in order to explain the 'Second Land Use Survey' of Great Britain. Coleman attempted to find out five categories of land use i.e., farm scape, urban scape, wild scape, urban fringe and marginal fringe[8] (Fig. 6.5). It is felt that the conventional land use types must have been of lower in hierarchical orders as the fundamental territories of land use are found continuously up to the marginal fringe. Such variations are found due to physical and socio-economic influences. The interacting zones and socio-economic influences. The interacting zones are

FIG. 6.5 : *General land use model (after Rawson and Sealey)*

found in advancing, stable and retreating stages. Further, it is hoped that these stages and the reasons of their formation should be really of wider application all over the world.

For a well structured general model, according to Coleman, the presence of the following characteristics should be a prerequisite i.e.,

(i) a land use model must be comprehensive,
(ii) the model must be integrated,
(iii) the model must be simple in order to give added insight,
(iv) the model shall be hierarchical,
(v) the model should be practical,
(vi) the model should be visual-spatial quality,
(vii) the model should be flexible, and
(viii) the model should be genuinely general in its application.

The comprehensiveness means that all categories of land use must be considered in the model as an integrated system. Such integrative system does not require the formation of land use categories in isolation but it should include the variable available in a 'land use territory'. The idea of added insight reflects that the model should be comprehensive and simple. Hierarchy denotes the varying objectives of the model building i.e., territorial, partial, sample, regional, principle, etc. It is also expected that the model should be problem oriented i.e., planning the area, classification of an area and that must have been of high level academic interest. The characteristics of visual-spatial equality requires the geographical explanation landscape territories. Flexibility prerequisite in the model so that the change in the present land use could be easily included in the model. Further, it should have wider application in any part of the inhabited earth surface.

DIAGRAMMATIC MODELS

With a view to compare the models of land utilization in relation to field patterns in the Middle Ganga valley Hiroshi Ishida presented several diagrammatic models.[9] The hypothesis is that a ring shaped territorial structure of rural settlements have inner, intermediate and outlying lands. Although, from one region to the other the variations in fertility are found according to the varying conditions of land improvement. Since long this has been the key criteria of land classification for the collection of revenue. The evolution of rural settlements and the growth of agricultural system in India are the products of these ring shaped arrangement of the systems of land use and field pattern. Examples of the zonal variations in the arrangement of land utilization are as follows, Fig. 6.6.

Models of land use types in Eastern Uttar Pradesh

Bara : means incircled land: outer fringe of the settlement, rich soil and in most cases vegetables are cultivated.

Goind : Intermediate land around the village, i.e., land between *Palo* and *bara*.

Palo : Outlying unmanured land around the village beyond the inner and intermediate zones, and is the least productive land.

Gauchar : Village common pasture.

Dhankhar : Paddy field.

Model of land use patterns around rural settlements of Patna District, Bihar

Dih : Highland around village site used for vegetable crops.

Dhankhar : Lowland, used both for *Kharif* and *Rabi* crops.

Tal : Outlying land, only *Rabi* crops are cultivated due to flooding in rainy season, kharif crops may not be cultivated.

Model of land use of the Punjab village after consolidation

Haveli : Settlement site.

Awal : First rate land.

Bara : (Richest soil), vegetables are cultivated.

Nyayee : (Very rich soil), vegetables are grown and newly brought under plough.

Keyot : (also kheyot) Periphery.

Keyot 1 : (rich soil) sugarcane, cotton, wheat and barley are cultivated.

Keyot 2 : gram, bajra and jowar are cultivated and the area of fallow is large.

PS : Pasture.

Phirni : Circular road.

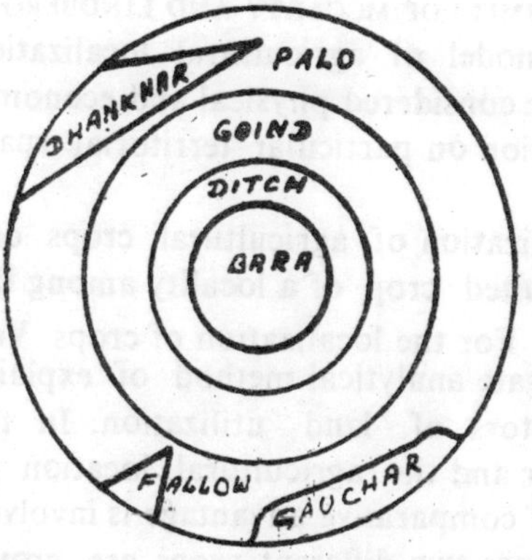

A

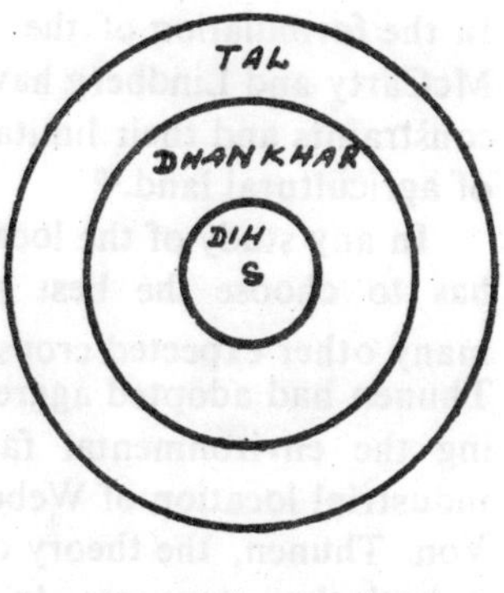

B

FIG. 6.6 : *(a) Model of land use in villages of Patna District, Bihar*

FIG. 6.6 : *(b) Model of land use in villages of eastern Uttar Pradesh*

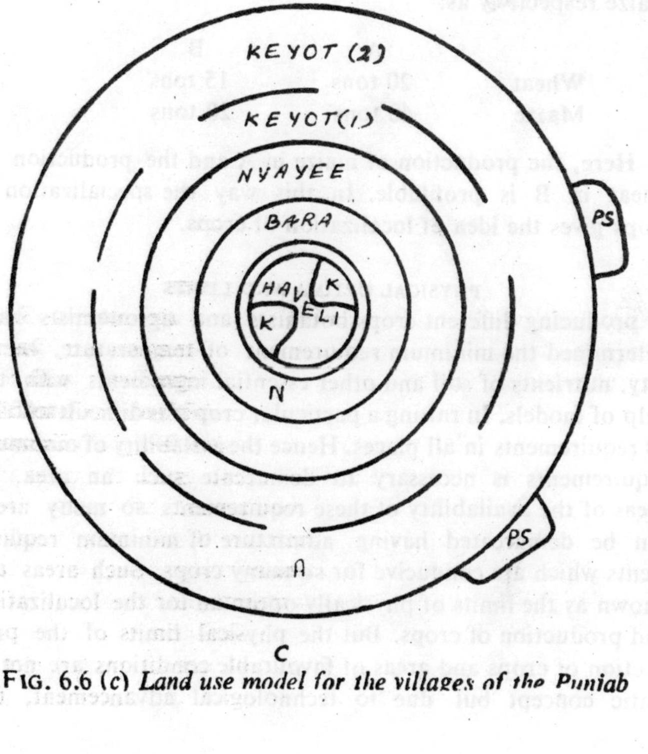

C

FIG. 6.6 (*c*) *Land use model for the villages of the Punjab*

MODELS OF OPTIMA AND LIMITS OF MCCARTY AND LINDBERG

In the formulation of the model of agricultural localization McCarty and Lindberg have considered physical and economic constraints and their limitation on particular territorial space of agricultural land.[10]

In any study of the localization of agricultural crops one has to choose the best suited crop of a locality among the many other expected crops. For the localization of crops Von Thunen had adopted aggregate analytical method of explaining the environmental factors of land utilization. In the industrial location of Weber and the agricultural location of Von Thunen, the theory of comparative advantage is involved in both these concepts. In case two different crops are grown in two different plots of land and in each plot only one crop could be profitably grown, than it is said that a particular plot has got specialization in producing a particular crop. Suppose A and B are two plots which produce wheat and Maize respectivly as:

	A	B
Wheat	20 tons	15 tons
Maize	40 tons	20 tons

Here, the production of maize in A and the production of wheat in B is profitable. In this way the specialization of crops gives the idea of localization of crops.

PHYSICAL OPTIMA AND LIMITS

In producing different crops botanists and agronomists have determined the minimum requirement of temperature, humidity, nutrients of soil and other essential ingredients with the help of models. In raising a particular crop it is difficult to fulfil all requirements in all places. Hence the avilability of minimum requirements is necessary to demarcate such an area. In areas of the availability of these requirements so many areas can be demarcated having admixture of minimum requirements which are conducive for so many crops. Such areas are known as the limits of physically optimum for the localization and production of crops. But the physical limits of the production of crops and areas of favourable conditions are not a static concept but due to technological advancement, the land resource, productivity of soil and changing inputs, the optimum physical limits for the productivity of crops are changing, e.g., with the technological advancement, now the dwarf variety of wheat could be grown in colder regions as well.

For the localization of agriculture in a region the required amount of temperature and humidity create favourable area for the production of crops. Because on this point less input and high output per acre of land are possible. From such centres, to outwards with the decreasing relationships of these two variables the production of crops are also decreasing and at the end there comes a line beyond which the production is impossible.

Figure 6.7 shows the physical optima and limits of cotton production in the cotton belt of U.S.A. This physical limit is determined by the amount of rainfall and time taken in the ripening of crops. But in cotton belt favourable conditions are found only in smaller areas where the fertile soils are available. Hence, the localization of agriculture is found only with

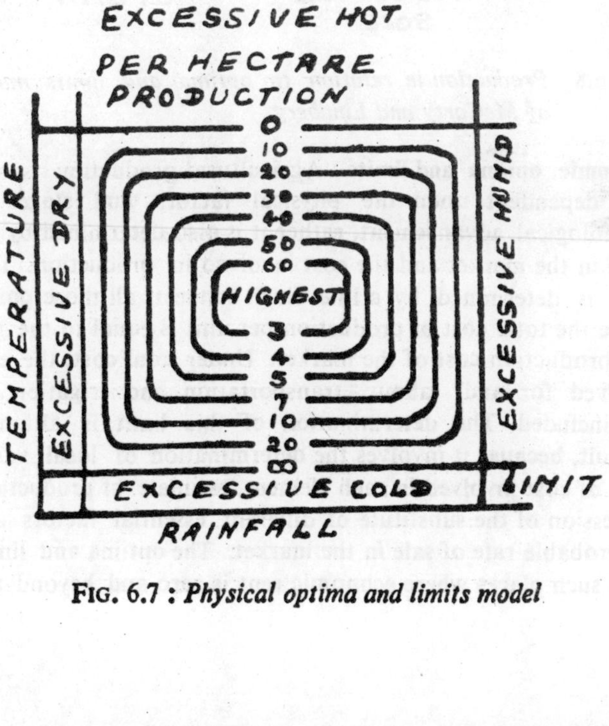

Fig. 6.7 : *Physical optima and limits model*

reference to physical optima and limits. In this way, the optimal physical limits can be determined for any crop in question, Fig. 6.8.

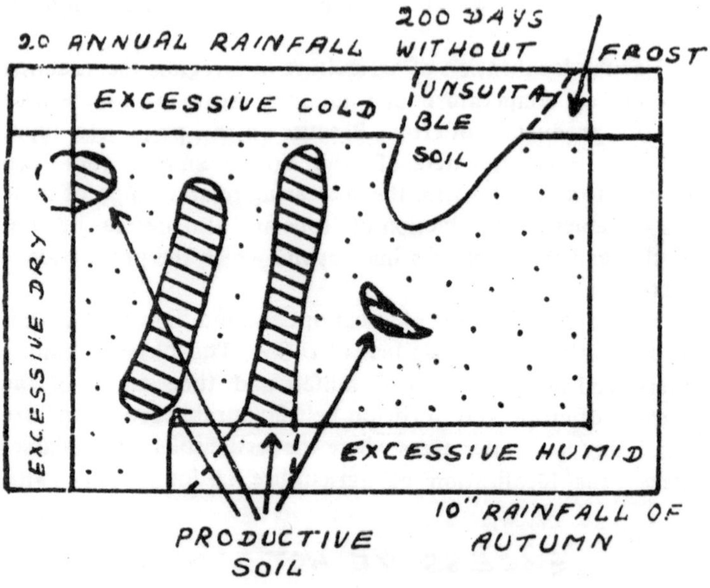

FIG. 6.8 : *Production in relation to optima and limits model of McCarty and Lindberg*

Economic optima and limits : Agricultural production is not only dependent upon the physical factors and stages of technological advancement, rather it is also determined by the price in the market and the cost involved in production. This limit is determined by a line which connects all those points where the total cost of production per unit is equal to the per unit production cost of the market. Under total cost, the cost involved for land, labour, transportation and irrigation are also included. The determination of this limit is although difficult, because it involves the determination of local variations of cost involved in each element, estimate of production, concession of the substitute of different essential factors and the probable rate of sale in the market. The optima and limits show such places where economic rent is zero and beyond this belt one has to observe disadvantage in production. Within this economic limit in any direction per acre production costs will be lesser and such areas may be available where in the market, per unit price and per unit total production costs will differ greatly. Such area is called an area of economically profitable condition Fig. 6.9.

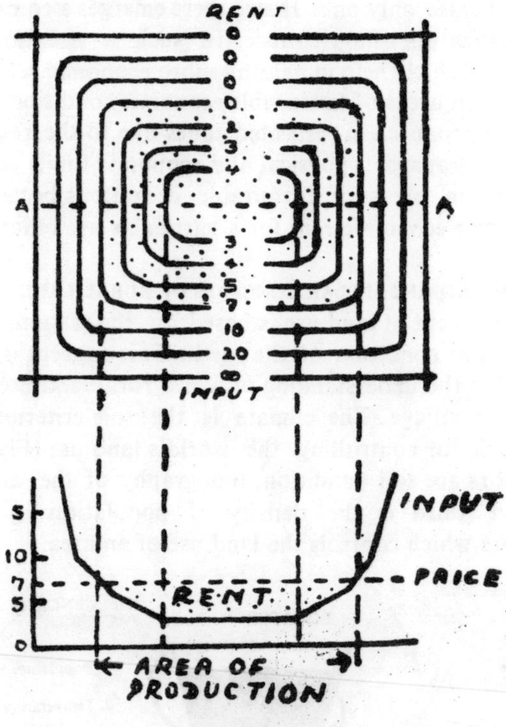

FIG. 6.9 : *Rent and price in relation to input on economic optima and limits*

Competition in between different crops : In the determination of favourable area for the economic limits of a crop, the probability of the production of other crops and the economic analysis of different crops could be made. In between two or more crops the cost-benefit analysis and its comparison with the economic rent could be easily made. In areas of favourable terrain and climatic conditions only such crops are grown which give per unit of area more production.

Quite often for a number of goods only one market is observed where the cost of investment is also the same. Usually market forms the areas of dense population, and for this fact for so many crops the favourable area of economic conditions is also only one. Hence there emerges a competition for the localization of agriculture. In such a situation from those crops which help in gaining more economic rent would be produced in areas of favourable economic conditions. Those crops which produced in restricted zones due to the restriction posed by the transport, physical and economic limits could be grown only in areas of favourable conditions because they provide higher economic rent for a particular use of land.

POLAR AND ANTI-POLAR CONCEPTS OF RURAL LAND USE

The polar concept of land use is based on the assumption of idealised polar condition of the earth. This concept is mainly used here for the understanding of world rural land use in the context of a village. The climate is the sole criterion which is responsible for controlling the world's land use (Fig. 6.10). Other factors are soil condition, topography of the land and slope and third is the density of population in certain temperature which controls the land use of an area.

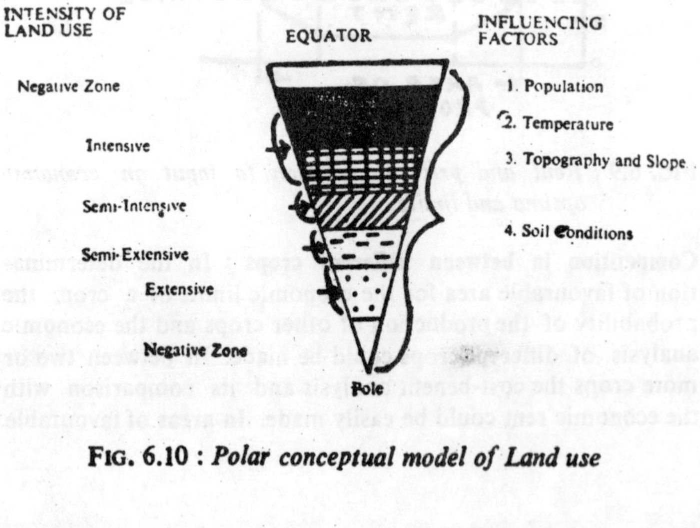

FIG. 6.10 : *Polar conceptual model of Land use*

This concept delineates that on polar region due to intense cold, soil becomes frozen with moisture, and secondly the barren land is also negative for human existence because unproductivity makes the land uninhabitable. The second zone is of lower land use because here the condition is improving than polar areas and few people live in this region after leading a wandering life based on prey of animals and fish in the lower land. The third zone of medium land use is characterised by lesser cold than the pole with improved soil conditions for agriculture and habitation of people. The inhabitants of this region are laborious and sturdy because they are always struggling with unsuitable environmental conditions for their existence. People of this region raised a single crop. The next zone is of high land use. Here because of timely rainfall and better soil fertility, the enrichment of humous from the leaves of forests, lesser amicable temperature and double crop pattern, the inhabitants of this temperate land have done a great miracle for the development of human civilization. The fifth zone is of higher land use with variety of crops, good rain and fertile soil which are favourable for the sustenance of life of greater number of people with 70° to 100°F temperature. This zone can be compared with that of the treble cropped land use zone of the former model (Fig. 6.10). The next zone is occupied by settlements which control the land use from the negative equatorial area to the distant polar regions according to their mode of accessibility and lesser security of crops and human life. Hence, this concept is known as polar concept of land use as propounded by the author. The only limiting factor for this concept is that there may notbe any physical barrier because highly rugged topographic feature may deviate the land use pattern.

ANTI-POLAR CONCEPT OF LAND USE AND RURAL SETTLEMENT LOCATION

On North and South poles of the earth the lines of meridians converge like spokes of a cycle and similar is the conditions of transportation lines which converge on the centre of the village. The consideration of the centre of the village as a pole is (Fig. 6.11) not appropriate, because the conditions in

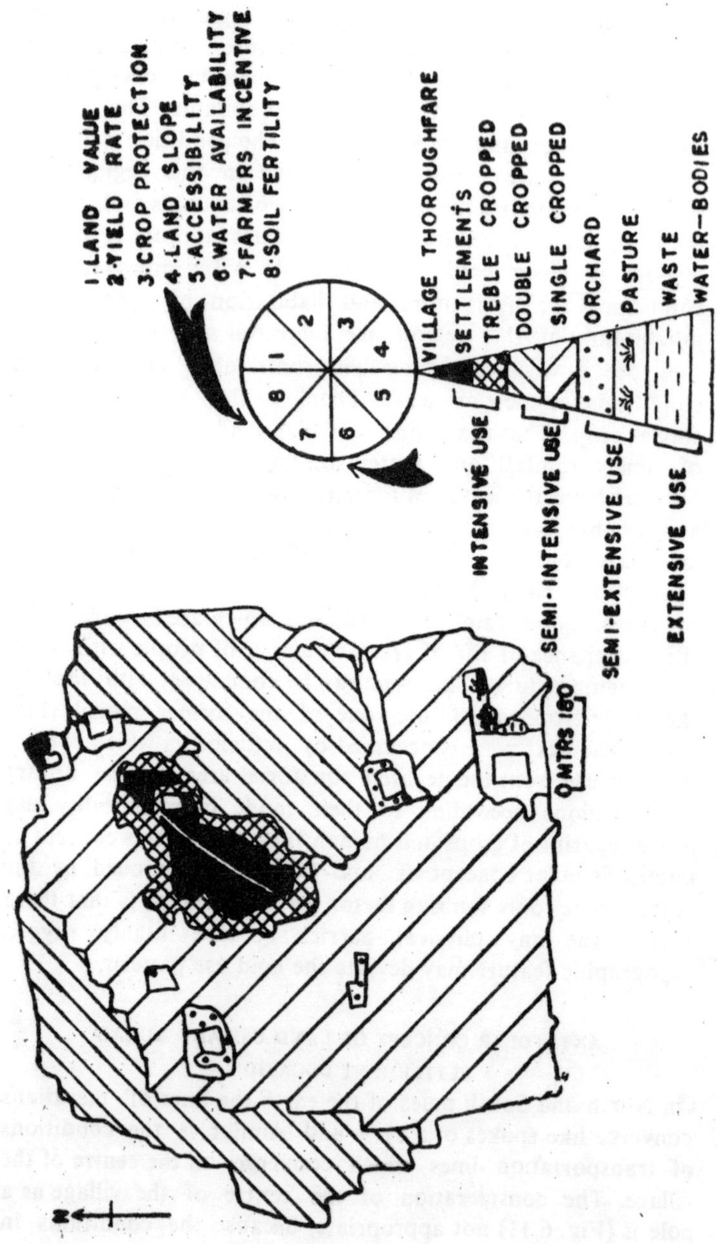

FIG. 6.11 : *Anti-polar conceptual model of Land use*

and around the village are not polar but anti-polar in overall conditions.

These polar and anti-polar concepts are controlled by several variables, whose intensity over landscape is higher near the settlement to high, medium, low and negative on the outskirts of the village, like that of the equator to polar regions of the world. These variables are price of the land, yield per acres, security of crops, topography of the land, slope, fertility of the soil, accessibility, availability of water and peasants incentive. It is termed as "anti-polar" concept of land use around a rural settlements.

This concept dictates that the intensity of land use is decreasing in the form of a series of concentric ring from the centre of the village to the outskirts. This is due to the higher input, better irrigation and treble crops of village *bariland*, while on the other hand, the features on outskrits of the village are single crop and less productive ground. This concept has lesser degree of similarity with Von Thunen's concept in varying assumptions. The extensive farming, increased area of cultivable land on outer margin is against the smaller areal extension of pole of the earth. Hence this concept rightly named as anti-polar concept of land use.

For the validity of this concept in Bihar, a village Bhalua of Saran district has been considered here. In this village, settlement area and treble cropped land have been considered as high land use, double cropped and single cropped land as medium land use, orchard and water bodies as low land use and others in the category of very low land use, beginning from the centre of the village to the outer margins.

In modern times, the scientific way of cultivation on the outer margins of the village suggests that now this concept is losing its ground. But this proves the validity of the concept that the previous condition of land use around the village was anti-polar type from the centre of the village to the outward and the same is still continuing for most of the villages in India. But nowadays, in some localities the facilities of irrigation, intensive cultivation, deeper harrowing of fields by tractor, use of chemical fertilizer and compost coupled with green manure on marginal fields are giving multiple crops and hence the security of crop is necessary.

This gives birth to smaller hutments and pumping set stations even outside the village, and this trend is universally discernible today. In this way, the dispersal of settlements is going on with the construction of flood protection embankment, and hence planned development of rural settlements and land use should be introduced by the government at this opportune moment of changing economy and the present state of living condition.

Statistical formulation of model: The land use intensity can also be in terms of the gravity model which is a measure of interaction of forces affecting particular use of the land and has been taken from Newton's law of gravitation. In this study the difference in the intensity of land use from the centre of the village to the outskrits is considered due to the spatial variation of the utilization of land. Hence the intensity of land use between the two places (*i* and *j*) in the product of the mass (*p*) of input of the two places divided by the distance (*d*) between them. This can be expressed as;

$$lij = \frac{Pi\ pj}{dij}$$

In course of time it can be hypothesized that the intensity of land use will increase in view of the increasing population and decreasing distance from the centre of the village to the outer margins. Thus in second stage the situation for the two places will be as follows:

$$lij = \frac{pi\ pj}{dij^b}$$

Thus the determination of land use from the centre of the village towards outer zone of *n*th places will be equal to:

$$\frac{pi\ pj}{dil^b} - \frac{pi\ p^2}{di2^b} - \frac{p\ p^3}{di3^b} - \cdots\cdots \frac{pi\ pn}{din^b}$$

$$= \frac{pi\ pj}{dij^b}$$

As *pi* is common throughout the whole summation, therefore, it can be expressed as

$$\frac{pi\ pj}{dij^b} = Vi(\text{Potential})$$

This division of land use intensity from the centre of the village to the extensive farming area has been designated as potential (V) for different zones or places. This may be calculated and the quality can be expressed as:

$$Vi = J = 1 \frac{pi}{pij^b}$$ for the number of land use patterns of different distance over space.

CONCLUSION

The generalization of the real world situation in a diagrammatic, mathematical and structural form is known as a model. In land use studies in each and every part of the world, the location of settlements in central part of the village surrounded by the vegetable gardens, agricultural fields and pastures are common with varying exceptions which are apparently related with the local physical, climatic, social, political and economic conditions etc. available in the region. This proves the validity of the modification of Von Thunen's model several times with some increasing scientific precision and regional ecological conditions. It is hoped that the 'Polar and Anti-polar' concepts of land use and the location of rural settlements' will help in understanding the problems of land utilization especially in rural areas. The better arrangement of the sectors of land in different parts of the village could be planned in a better way with the help of model development in this study in order to have a good harvest from farm with limited means of cultivation especially in Indian rural sectors.

REFERENCES

1. R.J. Chorley and P. Haggett, *Physical and Information Models in Geography*, Methuen, London, 1970, p. 22.
2. M. Chisholm, *Rural Settlement and Land Use*, Hutchinson University Library, London, 1962, pp. 21-35.
3. R.S. Thoman, E.C. Conkling and M.H. Yeats, *The Geography of Economic Activity*, McGraw Hill Book Company, New York, 1968, p. 153.
4. Ibid, p. 154.
5. S. Edgar and Dun Jr., *The Location of Agricultural Production*, University of Florida Press, Gainesville.

6. R.R. Renne, *Land Economics*, Harper and Brothers, New York 1947, p. 198.
7. A., Coleman, A geographical model for land use analysis, *Geography*, No. 242, Vol. 54, Part-1, 1969, pp. 43-55.
8. V.K., Shrivastava, General land use model: Method and experiment, *Uttar Bharat Bhoogol Patrika*, Vol. XIII, Nos. 1-2, 1977, p. 33.
9. H. Ishido, Village and field pattern in the middle ganga plain: diagrammatic model and some overseas comparisons, (Ed. R.L. Singh) *Rural Settlements in Monsoon Asia*, National Geographical Society of India, Varanasi, 1972, pp. 173-74.
10. H.H. McCarty, and J.B., Lindberg, *An Introduction to Economic Geography*, Prentice Hall, N.J. 1966.

7

Land Use Classification

LAND USE classification is the systematic arrangement of various classes of land on the basis of certain similar characteristics, mainly to identify and understand their fundamental utilities, intelligently and effectively in satisfying the needs of human society. Thus, land must be carefully utilized, so that it may fulfil our varied needs after its proper allocation. The best use of each parcel of land requires a scientific and methodically appreciable classification of the present land use. This may help us in investigating the land use problems and be the basis of planning for the best use of our land after considering the major land use categories.

The comparability of major land use categories shows that there is good relationship between forestry and grazing lands with the recreation, but there is a very poor correlation between wildilife sanctuaries with the agriculture and water reservoirs (Table 7.1). We may keep it under single versus multiple land use concept.

The land use types and its classification must be clearly presented in comparison with other land classification according to productivity index[1] and the yield and quality of crops grown under physically defined system of management or according to Storie index[2] based upon soil profile, soil texture and other physical factors combined to control the use capabilities of particular soil and its productivity under favourable environmental conditions. This use capabilities provide farmers a basis for producing over a longer period of time under stated conditions of use on specific parcels or units of land.[3]

TABLE 7.1 : Comparability of major land use

Primary Land Use	1	2	3	4	5	6	7	8	9
Urban Purposes	Complete								
Recreation	none	Complete							
Agriculture	none	very poor	Complete						
Forestry	none	high	none	Complete					
Grazing	none	high	none	very poor	Complete				
Transport	none	none	none	none	none	Complete			
Reservoirs	none	poor to high	very poor	very poor	poor to Moderate	none	Complete		
Wildlife Sanctuaries	none	high	very poor	Moderate	Moderate	none	poor	Complete	
Mineral Production	none	poor	poor	Fair	Fair to Moderate	Fair	poor	poor to Fair	Complete

Source : M. Clawson et al., *Land for the Future*, John Hopkins University Press, Baltimore, 1960.

OBJECTIVES OF LAND CLASSIFICATION

The purpose of land use classification is to maximise the productivity and conserve the land for posterity. A classification of land requires considerable time and expense, hence, there must be a justification to classify any land into different categories. The need is to specify clearly the objectives of land classification procedure which should be applicable all over the world with some difference or similar principles. "Land classification is not an end in itself but a means of obtaining better land use".[4] The objectives of classifying land may be grouped into eight major categories according to Barnes[5]: (1) More enlightened and economically sound land settlement policies both public and private, (2) guidance in public land purchases and development, (3) Planning the organization and distribution of local governmental services, (4) guidance in the distribution of public aid, particularly federal and state subventions and public relief to local areas, (5) guidance in determining sound real estate lending and borrowing policies; (6) equalization of land assessments for taxation purposes; (7) guidance in developing and administering programmes for improved land utilization practices, soil conservation and erosion control measures and similar practices, and (8) guidance in developing sound farm management policies and organizing the most effective type and size of operating units.

Further, the land use classification should also include the natural qualities of the land, economic considerations such as market accessibility, size and type of operating unit, distribution and composition of population; location of roads, schools, power lines, stores, factories, mines, types of properties and ownership, economic outlook, price prospects; transport facilities and the cost of production must be properly analysed in the classification of land before putting it into best possible use.

DEVELOPMENT OF LAND USE CLASSIFICATION ON REGIONAL LEVEL

This section is concerned with a purely inventory classification of land as to how much area is utilized by a particular type of use at the time of survey. After finishing the detailed land utilization survey, it is necessary for an investigator to classify the different uses of land into certain fixed categories which may be useful for all practical purposes and planning the future use of land in any region. The classification of land may be made at the international, national, state and village level of our usable land. This classification is simple enough to be understood and applied by the voluntary workers and students with a very simple training. In this regard, the research organizations and scholars have given the following different categories of land use classification.

TECHNIQUES OF LAND CLASSIFICATION

In 1871, Wheeler was directed to prepare topographic maps and classify the land use of the United States. The potential use of land was classified into four categories (a) land suited for agriculture; (b) land suited for timber; (c) land suited for grazing; and (d) arid land. In between 1874 and 1879 when the land use classification work was added to the U.S. Geological Survey, some 448,000 square kilometres were classified.[6]

UNIT AREA LAND CLASSIFICATION

W.D. Jones had returned from patagonia with a view to preparing land use maps on the same scale and degree of generalization as the maps of physical features. The basic problem involved in the land use classification of agricultural areas was how to identify and plot the significant units of area on maps relevant to the understanding and guidance of land use. Firstly a proposal was made to prepare separate maps on the same scale to show the critical elements in land use problems. Secondly, the number of such maps should be reduced e.g. one to show the conditions of the physical land including units of soil, slope, drainage, and cover of wild plants; and the other to show the categories of land use. When these two aspects of the same maps were superimposed, the precise covariance of the physical features and the land use could be examined in detail. "Later the suggestion was made that all this information could be plotted on one map by making use of a fractional code symbol." The denominator of the fraction was made of digits representing categories of soil, slope and drainage; and the numerator was made of digits standing for types of land use or wild vegetation (Table .2). In the course of field-mapping, when the observer noted any difference of the physical land or the land use, a boundary had to be drawn and a new fractional code symbol shown. In each unit area, represented by one fraction, there was a single, uniform association of land quality and land use. Of course, since every microscopic point on the face of the earth differs from all other points, these unit areas were in fact, generalizations. Within each unit area there was a certain range of diversity that had to be small enough to be acceptable and this, of course, depended on the proposed use for the resulting map.

The fraction code system was tested by Finch with a group of graduate students from the University of Wisconsin, and the results were published in the so-called Montfort Study in 1933 (Finch, 1933). The map was published on a scale of approximately 1/15000 Fig. 7.1 and provided an extraordinary amount of detailed information. From this map it was possible to prepare a number of thematic maps of individual elements such as slope, soil, or land used for specific purposes. But the field-mapping had required so much time that Finch was unable to recommend the method as useful.

132/11	112/11	112/11	X	213/11
142/11	121/11	171/11	X X	133/11
105/11	212/11	132/11	111/11	211/21
121/11	181/11	131/11	171/11	151/11

FIG. 7.1 : *Unit area method of land classification*

TABLE 7.2 : **Numerator**

Left-hand digit major use type	Second digit specific crop or use type	Third digit condition of crop
1. Tilled Land	1. Corn (Maize)	1. Good
	2. Oats	2. Medium
	3. Hay in Rotation	3. Poor
	4. Pasture in rotation	
	5. Barley	
	6. Wheat	
	7. Peas	
	8. Soya Beans	
	9. Potatoes	
	T. Tobacco	
	X Sudan grass	
	2/5 Oats and Barley mixed	
2. Permanent Grass land	1. Open grass pastures	1. Good
	2. Pasture with scattered trees or bush.	2. Medium
	3. Wooded pasture	3. Poor
	4. Permanent grass cut for hay	
3. Timber Land	1. Pastured	1. Good
	2. Not Pastured	2. Medium
		3. Poor
4. Idle Land	1. is capable of use	1. Good
		2. Medium
		3. Poor

Denominator

Left hand digit slope of land	Second digit soil type	Letter X conditions of drainage
1. Level 0°–3°	1. Marshall Silt loams	X Poor
2. Rolling 3°–9°	2. Knox silt loam	XX Very Poor
3. Rough 9°–15°	3. Lintonia Silt loam	
4. Steep, over 15°	4. Wabash silt loam	
	5. Rough stony land	

Source : P.E. James, All Possible World, Chapter 13, p. 414.

Further this method of land classification was developed by the Tennessee Valley Authority. In this method variables related to the physical and cultural environment are tabulated and numbered. Further the numbers for each unit area of land are expressed in a fractional code. The digits as indicated below in the denominator represent physical conditions and digits in the numerator stand for cultural environment. In case of change in a factor a new class is represented. The symbol for each unit of the land shown in three parts, e.g. a Roman numerical a short fraction and a long fraction as indicated below.

$$V\frac{2}{3} \quad \frac{IG\ 234}{2331123}$$

Where the Roman numericals from I to V show the seriousness or absence of problems in the area as I—represents insignificant; II—not critical; III—moderately critical; IV—very critical; V—so critical that a radical change in land use is desirable. The short fraction is an abstract of the long fraction. On numerator side areas are classified into five classes (1—5) i.e., excellent, good, medium, poor and very poor; for denominator, seven digits have been used in which each relates to a condition, i.e. 1—slope, 2—drainage; 3—erosion; 4—stoniness; 5—rock exposure; 6—soil depth and 7—soil fertility. These physical conditions in each case expressed with seven digits and is further classified into five groups in case of slope drainage, erosion, rock exposure and soil fertility, while stoneness and soil depth have four subdivisions. The applicability of the method depends upon availability of facilities for field work and aerial photographs, etc.

INDUCTIVE METHODS OF LAND CLASSIFICATION

Sometimes it is known as 'Storie Index Method' and sometimes Inductive Method. It is based on soil profile only. To measure the Storie index it compiles the rating from percentage values of the soil profiles. Three variables are considered. One for the general characters of the profiles; second, for the surface future and a third, for 'soil modifying conditions', e.g. drainage, acidity and erosion. The figures for those three factors expressed in percentages of the optimal conditions for plant growth and multiplied together to obtain the rating. The product is expressed as the percentage of the maximum. The significance of multiplying instead of adding the points induced for each soil characteristic is that only one abnormal factor can dominate the final rating. In this soil may have excellent profile conditions if it is rated at 100 for the 1st factor, excellent texture if rated at 100 for the second factor, but because of acidity or alkalinity it is rated at 40 for the third factor. The product of three factors gives a final percentage rating of the soil of 40 even if other conditions are favourable, the alkalinity renders the soil unproductive. On the basis of addition, the rating will be 80 per cent. The results may be represented mathematically which are as follows:

I Case : $\frac{100 \times 100 \times 40}{100 \times 100 \times 100} = 40$ per cent

II Case : $\frac{100+100+40}{100+100+100} = \frac{240}{300} = \frac{8}{10} = 80$ per cent

This method shows that inspite of subjective award of points the high degree of comparability can be made possible with the help of Storie index method.

DEDUCTIVE METHOD OF LAND CLASSIFICATION

The U.S. soil survey productivity ratings analysed the deductive type of land classification as follows:

TABLE 7.3 : **Type of land classification**

Soil type	Corn.		Wheat		Rye		Alfalfa		Sugar beet		Productivity Grade	
	A	B	A	B	A	B	A	B	A	B	A	B
Miani Loam	70	90	70	100	80	100	70	90	60	70	3	1

Source : G.V. Jacks: Land classification for land use planning', *Imperial Buraeu of Soil Science*, Technical Communication, No. 43, 1964, p. 69.

The values in column **A** and **B** refer to the percentage of the standard yield of crop (A) with the common practices of management in the area and (B) as the best current practice. The standard yield refers to average yields obtained without the use of any extra input.

The 'productivity grade' can be calculated with the help of simple percentage of the crop ratings according to each 'crops' local significance. The values 70 for (A) and 90 for (B) can be calculated as follows:

$$A=\frac{35\times 100}{50}=70$$

$$B=\frac{45\times 100}{50}=90$$

In the some way the values for A and B in the other columns can be calculated. The variation in between A and B ratings is a measure of the soil's response for management.

Productivity grade of soils indicates the general productivity for the crops of the region. Soils having an average value of 100 to 90 are graded 1st class, between 90 and 80 as 2nd class and so on. Low productivity grade would be classed as low in fertility. The determination of productivity is an arbitrary procedure which depends on the local significance of a land as alluvial soil of North Bihar is used predominantly for cereal crops, as its productivity grade corresponds to productivity ratings with respect to the cereals only.[7]

"This fractional code system implies the existence of a unit area—uniform with respect to the physical land and the land use. The field operations of the survey resulted in a series of maps of individual elements: lay of the land (showing five categories of shops); soil types (based on the standard definitions of the U.S. soil survey); drainage features; Cover (including wild vegetation, crops or planted pastures, and also abandoned farms); population; political organization; assessed valuation; tax delinquency; land ownership and where taxes were still being paid, the intention of the owner in maintaining possession of the land and trade areas. Each of these items was plotted separately on a map but their compilation in association show some repetition of the same items. Not only could certain natural land types be identified (repeated associations of slope, soil drainage and wild cover), but also certain economic conditions were found to have high correlation with certain land types".

WORLD LAND USE SURVEY AND CLASSIFICATION

The increase in population needs additional land for shelter and food produce and requires judicious utilization of our land resource. In view of this surging problem, world land use inventory survey had been proposed[8] in the International Geographical Congress at Lisbon in April, 1949. An overwhelming support was received from Unesco to finance this ambitious project. Later on a commission of five persons met at Clark University, Worcester, Massachussetts in Decemer, 1949, which had formulated a plan of objective and procedure to survey the world land use.

On this basis land use survey has been carried out in several countries including Poland, Cyprus, Italy, Jamaica and others. There is an urgent need to carry out these surveys; and the maps prepared afterwards should be compared. To coordinate the works of different world organizations is an another important task for the commission on International Land Use Survey. The task of comparing these maps had already begun by "Pan American Institute of Historical and Geographical Studies" covering the whole of America.

To accelerate this gigantic task, an office was also established in Royal Geographical Society, London in 1951 to correlate the works carried out by individual countries all over the world and to train the personnel to conduct the survey. Later on a permanent office of world land use survey was established there. It was proposed to collect the information in the field on the maps available locally and finally the publication of these maps was planned on a uniform scale of R.F. 1:1,000,000 and each map was to accompany an explanatory text. At the second time, the commission met at the International Geographical Congress in 1953 and it was proposed to carry out pilot survey in as many parts of the world as possible. L.D. Stamp was made the incharge of the Eastern Hemisphere and Van Valkenburg of the Americas. "An impressive record of Pilot survey on different scales of various parts of the world were presented".[9]

These Commissions proposed a simple classification of World Land use along with colour scheme which is mainly suited to local condition. The classification is as follows:

World Land Use Survey was drawn up under the auspices of Unesco:[10]

1. Settlements and associated non-agricultural land (dark and light red).
2. Horticulture (deep purple).
3. Tree and perennial crops (light purple).
4. Crop Land: Continual rotation cropping (dark brown), land rotation (light brown).
5. Improved permanent pasture (light green).
6. Unimproved grazing: used (orange), not used (yellow).
7. Woodlands: dense (dark green), open (medium green), scrub (olive green), swamp forests (blue green), cut or burnt over forest (green stipple), forest with subsidiary cultivation (green with brown dots).
8. Swamps and marshes (blue).
9. Unproductive land (grey).

It is impossible for a single organaization to carry out world land use and hence this task has been given to the respective countries of the world so that the usefulness of air photographs can be tested and the surveys carried out by different nations be readily comparable with one-another. The world land use survey commission rendered reports on its work at Washington Congress in 1952, the Rio Congress in 1956, Stockholm in 1960 and the progress has been noticed in several countries where land use planning is necessary for land effective use of our land resource. As for example, Pakistan had commissioned an aerial survey of the great area of the Indus Basin upon which the future production of foodstuffs and export of that country depend. Similarly, the Canadian Government arranged an aerial survey of the vast tract of forest-land, which are practically impossible to traverse from the ground and land use maps have been prepared for future use.

London headquarters of the world land use survey has carried out pioneer studies of land use in Africa, e.g. in Tanjanyika and Nyasaland, where the huge extent of bush, sparsely inhabited vast tracts of the country vary very little from one part of the country to the other.[11] The publication of world land use survey map is on the scale of 1:1,000,000, which is approximately 16 miles to 1 inch. It was found that the air photographs showed a most remarkably complex character of vegetation and land use in those areas.

TYPOLOGY OF AGRICULTURAL LAND UTILIZATION

With a view to classifying world land use Kostrowicki (1972,[12] 75,[13] 76[14]) and Beek and Bennema (1972)[15] have developed typology of agricultural land utilization. Kostrowicki aims at providing world typology of agricultural land use following the scheme of world land use survey. Kostrowicki has considered a number of variables in a way to develop typology of agricultural land use which are given in Table 7.4.

TABLE 7.4 : Selected diagnostic features and variables for the typology of world agriculture

No.	Variables	Expressed by
Social and Ownership Characteristics		
1.	Land ownership	A—E Symbols.
2.	Land operation	A—E Symbols
3.	Size of holdings	1—5 classes
Organizational and Technical Characteristics		
4.	Inputs of labour	1—5 classes
5.	Inputs of animal power	1—5 classes
6.	Inputs of mechanical power	1—5 classes
7.	Organic manuring	1—5 classes
8.	Chemical fertilizing	1—5 classes
9.	Extent of irrigation	1—5 classes
10.	System of irrigation	A—E symbols
11.	System of land use	A—E symbols
12.	System of crop or land rotation	A—E symbols
13.	Intensity of land use	1—5 classes
14.	Cropping system	1—5 classes
15.	System of livestock breeding	1—5 classes
16.	Permanent grasslands	1—5 classes

Production Characteristics

17.	Land productivity	1—5 classes
18.	Labour productivity	1—5 classes
19.	Level of commercialization	1—5 classes
20.	Degree of commercialization	1—5 classes
21.	Degree of specialization	1—5 classes
Structural Characteristics		
22.	Production orientation	1—5 classes
23.	Orientation of commercial production	1—5 classes

Source : J. Kostrowicki, 1975.

The use of variables in the above table leads to a system of 'model types of world agriculture', which is represented by a formula in which the various factors have given rise to different types of land use.

"Agriculture, whether in an agricultural holding or in another unit of study, has, therefore, been presented by the following code:

$$T=S\frac{O}{P}C$$

in which T means the tpye of agriculture, S=social and ownership characteristics, P=production characteristics, and C= structural characteristics or enterprise combination".[16]

With the help of the above formula 53 model types have been developed which are grouped into a number of main types:

Primitive Agriculture

1. Nomadic herding
2. Shifting (long fallow) agriculture

Traditional Agriculture

3. Current fallow agriculture
4. Continuing, extensive, mixed agriculture
5. Labour intensive, non-irrigated, crop agriculture
6. Labour intensive, irrigated crop agriculture
7. Labour intensive, irrigated, semi-commercial crop agriculture

8. Labour intensive, non-irrigated, semi-commercial crop agriculture
9. Low intensive, semi-commercial crop agriculture
10. Large scale, low intensive, semi-commercial agriculture, latifundia

Market Oriented Agriculture

11. Mixed agriculture
12. Intensive Agriculture with fruit growing or market gardening dominant
13. Large scale, specialized agriculture with livestock breeding dominant
14. Plantation agriculture
15. Specialized irrigated agriculture
16. Specialized, large-scale grain crop agriculture
17. Specialized large scale grazing

Socialized Agriculture

18. Mixed agriculture
19. Specialized fruit or vegetable agriculture
20. Specialized industrial crop agriculture
21. Specialized grain crop agriculture
22. Specialized grazing
23. Labour intensive, non-irrigated crop agriculture
24. Labour intensive, irrigated crop agriculture

In the above classification each type is divided into several sub-types. Kostrowicki has also given indications of the regional occurrences of various types. The characterization is also made by the construction of 'typograms' for each land utilization type. Here as an example typograms of 'nomadic herding' is given below:

1. Nomadic herding

$$C, C, 3\text{-}4, 0, 3\text{-}4\text{-}2\text{-}3 \quad \frac{1, 1, 0/1, 0/1, 0/1, 0/2\text{-}3, 0/1, 5, 1\text{-}2}{1, 1\text{-}2, 1\text{-}2, 1\text{-}2, 1\text{-}2} \; 4\text{-}5, 5$$

Where, "Common migratory grazing of herd animals, on common pastures (*C,C*), by larger or smaller (3-4) groups of people (tribe, clan or other communities) with or without some land cultivation, size of holdings in terms of acreage not identifiable (0). Number of animals per group of people medium to large (3-4), gross production small to medium (2-3). Labour and capital inputs per unit of area very low (1). When some land is cultivated, it usually happens with shorter or longer fallowing (2-3). Permanent grass lands dominate in the use of land (5), Density of livestock very low to low (1-2). Land productivity very low (1), labour productivity and commercialization very low to low (1-2). Degree of specialization very low to low (1-2). Livestock products dominate in gross (4-5) and commercial production (5). North African, the Middle East".[17]

BRITISH LAND USE CLASSIFICATION

First approach : In Great Britain, after a long and arduous field work, the first workable and satisfactory land use classification was made in the autumn 1930's-37, under the direction of L.D. Stamp. The basic classification for mapping on the scale of 1 inch to a mile along with a colour scheme was as follows:[18]

1. Forest and Woodland (dark green);
2. Arable land (brown);
3. Meadow land and Permanent grass (light green);
4. Heath and Moorland (yellow);
5. Houses with gardens, orchards, nurseries (purple);
6. Land agriculturally unproductive, close building, pits, quarries, cemeteries and industrial premises (red).

This classification also includes various symbols, shading and stipples in block to indicate certain sub-divisions.

Second Approach : With the passage of time, after Stamp's Survey of 1930, a recasting of the above scheme began to be felt necessary, and the Second Land Use Survey[19] was initiated in 1960's. The idea was jointly developed in a handbook by Miss Coleman and Mr. K.R.A. Maggs. This new classification was a revision of the old Land Utilization Survey Scheme and was more elaborate and largely extended in scope and it included other categories of use also. The sheets are published on the 1:25,000 scale. In her classification,[20] "grass land includes fields with scrub; arable land contains fields of cereals, by legumes, roots, green fodder crops, industrial crops and fallow; market gardening includes nurseries, flowers, fruits and vegetables. In the category of orchards, investigators were asked to record the land under grass, arable cropping or market gardening. The use of land under urban areas are sub-divided into commercial and residential houses with gardens, newly built up areas and caravan sites. The land under industrial category is sub-divided into manufacturing, extractive, tips and public, utilities". Investigators were also asked to include land under crops and vegetation of heathland, moorland and other rough land. This survey showed the land use changes in comparison with 1930's survey and the investigators were asked to collect the information on minutest level, so this added more skill and refinement to the collection of data and land use mapping.

Third approach : L.D. Stamp had suggested the classification of the land of Britain into categories, for broad national policy of land use planning and conservation of land resources. He had proposed three major categories and 10 types, based on:[21]

(a) the nature of the site (elevation and slope).
(b) the nature of the soil (its depth, texture and water conditions).

The major categories are as follows:[22]

1. Good, 2. Medium, and 3. Poor.

The ten sub-types are:

Good quality land :	1. First class land,
	2. Good general purpose farmland,
	3. First class land with grass,
	4. Good but heavy land,
	5. Medium quality light land.
Medium quality land :	6. Medium quality general,
	7. Poor quality heavy land,
	8. Poor quality Mountain and Moorland,
Poor quality land :	9. Poor quality light land,
	10. Poorest land,

First class land : It is intensively cultivated particularly for the production of food-stuffs. The soils are deep loams but it also includes some peat, fine sand, silt and loamy clay. Drainage should be moderately free and soils should not be stony so that it must be workable in all seasons.

Good general purpose farmland: It is similar to the first category but varies with respect to (a) less depth of soil, or (b) presence of stones or (c) occasional liability to drought or wetness, or (d) some limitation of seasons when the soil works easily, resulting in the restriction of a range of usefulness.The land is suitable for arable farming.

First class land with grass : The terrain condition is similar to the first category but it varies in relation to (a) a high permanant water-table, or (b) liability to winter or occasional flooding, or (c) somewhat heavier or less tractable soils. It is less suitable for arable cultivation than grass.

Good but heavy land : In this category the soils are heavy clays and heavy loams with the result that both the period of working and the range of possible crops arc restricted. The Boulder clay of Scotland represents this quality of land.

Medium quality light land : It is characterized by light soils having shallow layers. This includes (a) shallow light soils on chalk or Jurassic limestones, (b) shallow soils of older limestones and (c) light soils including gravels.

Medium quality general purpose farm land : This quality of land is designed to be defective due to steep slope and undulating land with varied water conditions. Hence the soil is defective by reason of stoniness, shallowness, heaviness, which helps in raising a mixture of crops and grasses.

Poor quality heavy land : It comprises clay lands and low-lying areas of undrained mosses which requires extensive drainage works before they can be rendered agriculturally useful.

Poor quality mountain and moorland : This quality of land is covered by varied character of the natural or semi-natural vegetation.

Poor quality light land : It comprises overdrained lands overlying coarse sands or porous gravels which include both coastal sand dunes and the inland sandy 'wastes' or heathlands.

Poorest land : In its present state, this land may be agriculturally useless but the possibilities of reclamation is always there. Salt marshes can be drained or site may be raised or sand dunes can be fixed soon.

LAND USE CLASSIFICATION IN THE U.S.A.

First approach : In the United States Land classification is linked with resource base with a desire to promote their wise use. Hence, many states have undertaken land classification work to meet the current land use problems. In these states, land capability classifications have been developed in determining land use adjustment programmes for the best suited use of land resources. There were wide variations in the techniques of land classification in various States and in each case land capability classifications had been made according to the situations and land use problems of a particular region.

In the U.S.A. the widely used schemes of land use classification are based on land capability classes. Three classes are generally determined which are based on the local relief, climate, soil and vegetation. The land is then classified in three categories according to anticipated results of different types of use:[23]

"A. Land which is cultivable but ranges widely in fertility —capable of carrying various crops or grass in rotation or permanent.

B. Land which should not be cultivated but is suitable for the poorer grades of permanent pasture (rough grazing) or timber production.

C. Land unsuitable for economic use as arable, grazing or afforestation."

The same method was used in the Montana Land Use Survey where a dozen "range types", ranging from grass land through bush and forest to absolutely bare ground were distinguished. A 'range type' is defined as 'a type of country which, because of a certain uniformity of climate, soil and vegetation may be submitted to uniform land use'. A modification to this scheme was presented for the aerial land classification survey commissioned by the Government of Jordan. Based on the same basis Government of Pakistan had prepared land use types from air photographs, covering the whole of Indus Valley Plain which was called a 'Soil Reconnaisance Survey'.

The soil conservation service and the agricultural land use planning activities of the U.S. Department of Agriculture had classified the land into capability classes,[24] which have been based mostly on the basis of the topography of the land. The following is the classification :

Class I : Very good land that can be cultivated safely and easily with ordinary farming methods.

Class II : Land that can be cultivated safely with moderate conservation treatments. These soils may be slightly erodible or may have water or climate problems.

Class III : Soil with considerable limitation in use and that requires intensive conservation treatments. Erosion, aridity, excessive wetness, overdrinage and salinity may be the causes of the problem.

Class IV : Soils that are severely limited in use. They can be cultivated only occasionally and with extreme care. These soils may be erodible, droughty, wet, overdrained or saline, so that the kinds of cultivated crops that can be grown as well as the number of years favourable for crop production are very limited.

Second approach : The Nantional Resource Planning Board on Land Classification in the United States, classified the land into five categories,[25] which is largely the result of the point of view of scope of the report:

Type 1 : Land classification in Terms of Inherent Characteristics.

Type 2 : Land classification in Terms of Present Use.

Type 3 : Land classification in Terms of Use Capabilities.

Type 4 : Land classification in Terms of Recommended Use.

Type 5 : Land classification in Terms of Programme Effectuation.

This classification of land is based on the natural qualities of land.

Type—1 : Land classification in terms of inherent characteristics : This is also known as the physical characteristics of land in which relief, soil characteristics, slopes, minerals, vegetation cover, rainfall, chemical composition of the land, climate, location of the land use site and the similar other physical factors should be taken into consideration. This requires field observation and mapping of the natural land qualities.

Type—2 : Land classification in terms of present use : Present land use is expressed in terms of some major land use i.e., agriculture, forestry, recreation, settlement and transportation. In some cases, this also includes statistical data dealing with relevant economic, social and political importance. Present use is again divided into (a) kind of use (b) characteristics of use, (c) Status of occupancy.

In 'kind of use' classification, the land is classified area by area, field by field to show how it is used. In area 'characteristics of use', the agricultural land use, types and yield of crops are mapped. In 'status of occupancy' collection of data related to land ownership mortgages indebtedness, tax delinquency, assessed valuation and related information are considered.

Type—3 : Land classification in terms of use capabilities : This classification is objective in its treatment, predicting what can be expected of the land if it is used in one way or other. "The use capabilities of rough, mountainous land, for example may be appraised and classified from the standpoint of forestry or recreation, or as a catchment basin for a municipal water-supply. In terms of forestry the land might be rated as poor, in terms of recreation as excellent, and in terms of water-supply as medium. Whether the same body of land is classified from one point of view or from several others, and whether the same or different natural qualities of the land are appraised in each case, the use capability map in terms of one use is not inter-changeable with the use capability map in terms of any other use".[26]

Type—4 : Land classification in terms of recommended use : Recommended use includes the physical characteristics of each parcel of land together with its present use and its capability in terms of every use to which land might be put. This use is based on constant effort to develop new methods and practice of land utilization.

A map of recommended use in a national forest, for example, is an expression of well-matured technique tempered by years managerial experience. The same holds good in case of recommended land use classification as practised in the development of an irrigation project. The resulting maps recommended that water be delivered to some bodies of land and not to others.

Type—5 : Land classification in terms of programme effectuation : This classification contemplates, some form of action, which puts desired uses into effect. In this way some modification and change in existing land use may occur. Thus, the form of map show that for each body of land when and how the effectuation of land use recommendation shall take place. The cost of contemplation uncertainty of availability of funds and the decisions may rest on the availability of materials. One example is as follows: Suppose the seedlings of paddy are ready for transplantation but, on the other hand, owing to excessive erosion in the transplantation it is not possible to transplant the seedlings without constructing embankment. The construction of side embankment may take one-year-time and at the time of completion of embankment the same seedling may not be there and this constitutes the management plan of land classification in terms of programme effectuation.

LAND CLASSIFICATION IN JAPAN

Generally the land utilization of Japan is classified into rice land, high land with dry agriculture besides this mulberry cultivation land, forested land and uncultivated land (Fig. 7.2). In Japan several land classification maps and their explanatory notes have been published in the last fifteen years. Most of them are mapped on topographic scale (1:25,000 and 1:50,000) and there are classifications of land forms, geology, soils and land use respectively with their applications.

Land use survey in Japan has been conducted since 1953 and has published 382 sheets on 1:50,000 scales, 19 sheets on 1:25,000 scales and 12 sheets on 1:200,000 scales (in Hokkaido) up to the present.

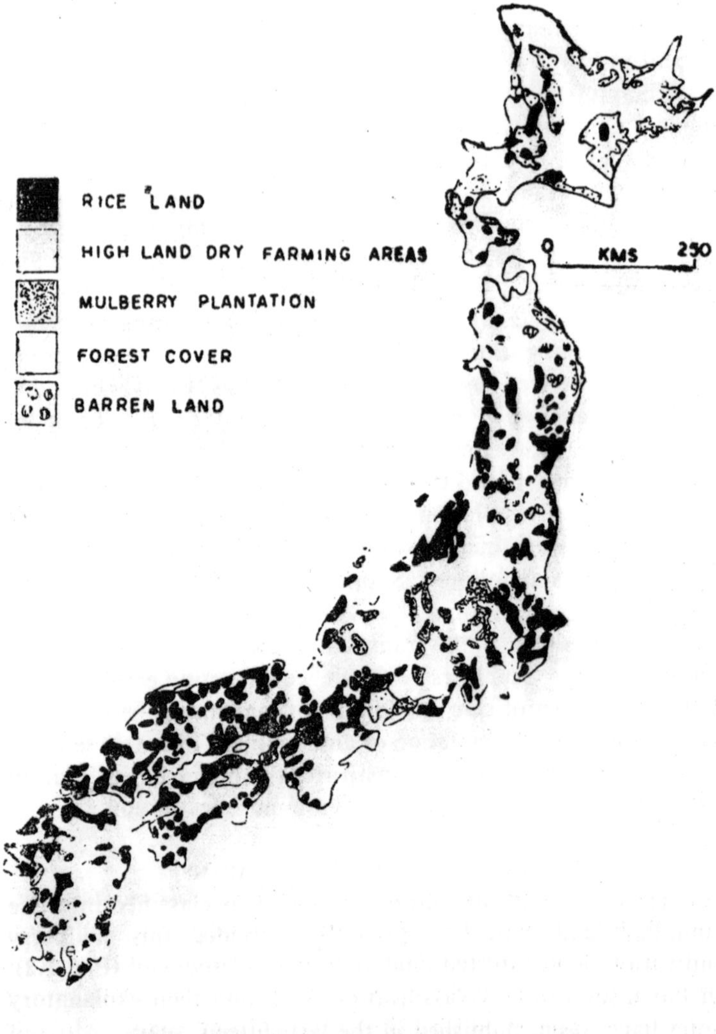

FIG. 7.2 : *Classes of Land use in Japan*

Classification is as follows :

1. Paddy Area has been shown by yellow colour and sub-divided into (a) triple cropped (b) double cropped (c) single cropped, (d) perennial cropped.

2. Upland field has been sub-divided into (a) triple cropped, (b) double cropped, (c) three crops every two years, (d) single cropped, (e) Fire field.

No symbol for main food crops has been used e.g. for rice, wheat and potatoes but for minor crops symbols have been used. They are as follows—M—Vegetables, X—Japanese peppermint, Q—sugar beets,—Asparagus; V--Grapes, T—Nursery, (x) Pyrethrum, 11 flax, O—forage crops, O—Apples and σ peas.

3. *Tree crops* : It has been shown by red colour and sub-divided into :

(a) orchards, (b) Mulbery gardens, (c) Tea gardens and other tree crops. (Symbols for fruit trees have been noted earlier).

4. *Forest area* : It has been shown by blue colour and sub-divided into:

(a) Conifers—Deep blue, (b) Broad leaf trees—Blue horizontally lined, (c) Mixed coniferous and broad leaf—Blue horizontally and vertically lined, (d) Bamboo, (e) Thin forest, (f) Deforested area, (g) Newly afforested area, (h) desert, trees and shrubs, (i) Area non-forested vertically lined. Different symbols have been used for different type of trees which are as follows:

S—Japanese Ceder and red wood, Ma—Japanese pine, K—Japanese larch, N—Japanese Oak, Kb—Birch tree, Ka—Oak, H—False arboritae, M—Yaro spruce, T—Momi (Fir) (Japanese hemlock), Si—Japanese linden, Ta—Ash, I—Painted Maple.

5. *Grass land* : It has been sub-divided into green (a) permanent pasture land which has been shown by deep green colour, (b) wild grass land—yellow coloured and green lened, (c) Water weeds, (d) Dwarf Bamboo, (E) Meadow, (F) Grass.

6. *Others* : It has been indicated by : (a) salt bed, (b) Barren land (c) Fish culture area, (d) Boundary of pasture (e) Boundary of seasonal grazing land, (f) Boundary of national forest, (g) Boundary of National Park.

7. *City and Town* : (a) Commercial Area, (b) Heavy Industrial Area (c) Light industrial Area, (d) Housing Area (e) Fishery Area, (f) Transportation Area (g) Public Area, (h) Rural Settlement.

8. *Transportation* : (a) Bus route, (b) Main auto rote, (c) Rail road, (d) Electric Railroad, (e) Special road for forestry, mining, etc. (f) Ropeway.

9. *Land Improvement and land conservation facilities :* (a) Boundary of well drained arable land, (b) Boundary of newly re-adjusted arable land (c) Boundary of terraced field area (d) Boundary of Area under reclamation, (e) Boundary of authorised soil conservation area, (f) Reserved forest area. (g) Main irrigation canals, (h) Level, (i) Revetment (j) cemented river bed, (k) Jetty, (l) Dam for power plant use (m) Dam of irrigation use (n) Dam for soil conservation and water level control use, (o) Dam for multi-purpose use, (p) Pier, (Q) Break water.

10. *Special facilities* : (a) Garden Frame (b) mine, (c) Power plant, (d) Transformer sub-station, (e) water level observatory, (f) water pipe.

11. *Land Features* : (a) Swampy land (b) Peat Bog.

In the above paragraphs symbols and classification are presented to represent the characteristics of complicated Japanese land use, which was once submitted to the Commission on World Land Use Survey of International Geographical Union. Agencies have been mostly utilized to take aerial photographs. Reconnaisance results of the photo-interpretation of land use is checked and supplemented in the fields. In Hokkindo where the pattern of land use is larger and less intricate, 1:200,000 scale land use maps have been compiled by areal photographic interpretation based on the data compiled by aerial photographic interpretation based on the data compiled as 1:50,000 topographic map of local government people.

Besides these land use maps compiled by G.S.I. which emphasized on the urban function have been compiled by the city planning authorities in several cities such as Tokyo, Kyoto, etc. Among them 1:30,000 land use map in the city portion of Tokyo showing the classification of urban function is much appreciated.

It can be concluded that land in Japan reflecting minor and complicated topography, dense population and long history has intensively utilized and very little land is left for natural vegetation. Consequently, project for better land use in Japan is the most important task for city planning and regional planning. To grasp the present status of land use in most cases it has been the necessity to survey the physical properties in case of city planning and regional planning project for better cultivation and plantation in agronomy, forestry and planning for housing, location of residential and industrial area.

Land use survey in Japan though not satisfactory, is relatively well conducted. It covers wide regions at comparatively high pace in surveying and compiling which considered to be based on the recognition about the above-mentioned necessity, supported by the enforcement of law of National Multipurpose Development, law of New Industrial City promotion besides wide diffusion of interpretation technique of aerial photography.

LAND USE CLASSIFICATION IN U.S.S.R.

The basic content of the general land use maps are the land distribution, characteristics of their present and future uses. It was (1:1000,000) that was accepted as a basis by the international committee on land use classification. This classification was refined and extended with the consideration of the U.S.S.R. specialist:

In view of the scales the following types of lands are distinguisked in the general land use map.

1. The lands belonging to the urban and rural localities to the industrial and transport departments,
2. Arable lands.
3. Perennial plantations.
4. Natural grasslands,
5. Forests,
6. Bogs and peat-bogs,
7. Unusable lands,

1. The lands belonging to the urban and rural localities are represented with a distinctive colour. Other lands of this group are shown in the maps either by means of line or symbols (roads, pipelines) or with the dash symbols (working and output of building materials, etc.).

2. This group contains the arable fields under cultivation, fallow, kitchen gardens, virgin and long-fallow lands. In the regions of irrigated agriculture the arable lands are subdivided into the irrigated and non-irrigated ones.

In maps, the virgin and long-fallow lands are shown with a distinctive colour, scale and the supplementary quantitative dashed symbols.

3. This group contains the garden, Vineyard, berry fields, the plantation of tea, citric plant and the mulberry. These lands occupy small areas and are wedged as parcels into other lands. Therefore, the areas occupied by the perennial plantations are painted with a single colour while the individual types are shown by supplementary symbols.

4. The natural grass lands are subdivided into the hayings and pastures. The flood lands have meadow haying. Dry valley haying are divided into the plain, foot hill, and alpine according to their relief features, and pastures with their seasonal uses.

In many areas of the U.S.S.R., the hayings and pastures are used for the same purposes. The meadows are used for cattle pasturage after the hay harvesting is completed thereon, while the hay harvesting is made on the pastures.

Vast areas of the north-eastern part of the U.S.S.R. are occupied by the deer pastures differing from other types of natural grass lands according to their quality and use. The deer pastures are painted in the maps with single colour.

5. The forests are sub-divided into the conifers, broad-leaved and mixed ones in their species, structure, wood procurement, processing and forest pastures.

6. The bogs and peat-bogs are sub-divided into the mossy, grassy etc.

7. Unusable Lands: include snow-covered areas, hill-slopes and summits.

LAND USE CLASSIFICATION IN CHINA

J.L. Buck, in his monumental study of *Land Utilization in China*,[27] concludes, from a survey of 16,786 farms in 168 localities of eight agricultural regions, that for agricultural China there can be no great increase in amount of farm land. The estimates that an additional 10 per cent of the present area in farms at most 9,311,740 hectares for the eight agricultural regions, consisting of 93,927,125 hectares of cultivated land could be made available for profitable cultivation by the following means: removal of graves from farm land, elimination of land in boundaries after consolidation of fragmented holdings, and putting under the plow arable lands not now cultivated. According to Buck, the arable uncultivated area amounts to about 14,170,040 hectares or 11 per cent of all uncultivated land of China.

Table 7.5 based on estimates from various sources, indicates that at present China's land suitable for farming is only about 15 per cent of its total land area and that one-fourth of the land area presently classified as arable is still uncultivated. More than 60 per cent of the land is either suitable for afforestation or utilized as grazing land. This shows that there is an ample scope for the extension of cultivation in China with the use of fertilizers, management of irrigation systems and the introduction of scientific and intensive cultivation with the help of developed implements.

TABLE 7.5 : **Land utilization in China**

Types	000' Hectares	% of total
Arable Land	147,400	15.28
Cultivated	110,550	11.46
Uncultivated	36,850	3.82
Forest Area	76,380	7.92
Land Suitable for Afforestation	201,500	31.25
Grass Land	268,000	27.78
Deserts, etc.	171,520	17.77
Total	964,800	100.00

Source : Food and Agriculture Organization of the United Nations, *FAO Production Year Book*, 1969, Vol. XXIII, Rome, 1970, p. 6.

Table 7.6 shows the classification of land of China according to elevation. It has been felt that area below 500 metres are suitable for farming. Areas between 500 and 1000 metres are hilly, but agriculture can be operated profitably in these areas with the help of terracing and other water and soil conservation devices. Areas between 1000 to 2000 metres high are mostly plateaus where animal husbandry is more profitable than farming. The bulk of the remaining land consists of mountainous areas with very little agricultural productivity.

TABLE 7.6 : **Classification of land of China according to elevation**

Elevation (Metres)	Area in 000' km^2	Percentage of Total Land Area
0-500	1,344	14
500—1,000	1,728	18
1,000—2,000	3,360	35
2,000—3,000	768	8
3,000—4,000	480	5
4,000—5,000	384	4
Over—5,000	1,536	16
Total	9,600	100

Source : Chang Lun. PIo, China's Resources, Peking, 1947, p. 1.

In this way, only 14 per cent of China's land is suitable for farming. Inclusive of 18 per cent 500-1,000 metres of high land, the total areas suitable for farming is still less than 33 per cent of China's total land area. Moreover, a substantial portion of the land under 1,000 metres of elevation is located in areas where climate and soils are not favourable for farming, due to uneven distribution of rainfall, short growing season, and poor soils. In China most of the arable land considered suitable for cultivation lies in Manchuria and Sinkiang, where a number of obstacles must be overcome before any large-scale reclamation programe can be carried out effectively. Most of the areas lie near frontiers, which are quite away from the trade routes and ports. Hence in order to use the land properly; agricultural developement must be accompanied by transportation, housing, public health facilities, industry and generation of electric-power.

Thus on the basis of above discussion A.K. Philbrick has classified the land use of China into the following categories (Fig. 7.3).

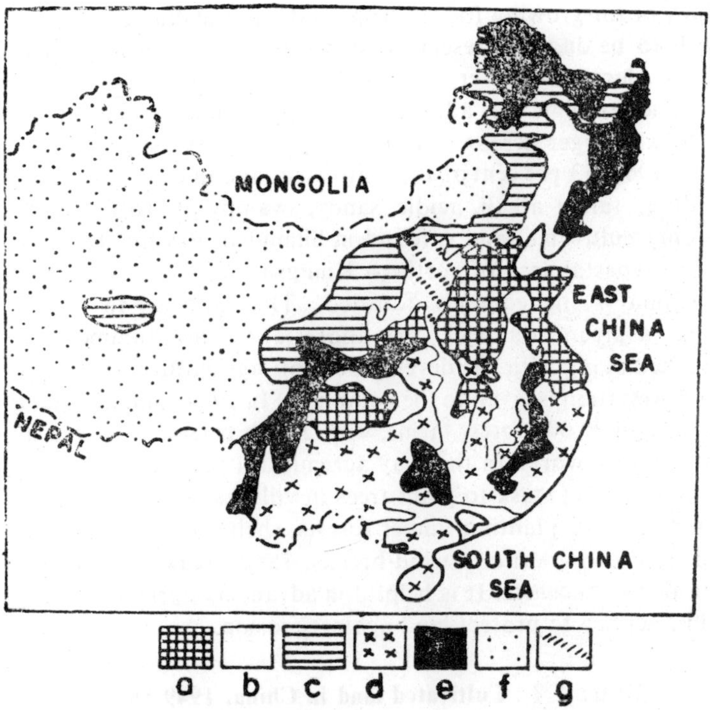

FIG. 7.3 : *Classes of land use in China*

a. Agricultural land 60 per cent or more under cultivation
b. Agricultural land at least 20 per cent under cultivation.
c. Advancing agriculture and afforestation.
d. Upland with some cultivation and patches of forest.
e. Natural forest.
f. Steppe and Desert grazing lands, some cultivation and patches of forest.
g. Major drought areas, 1960-61.

In China the 'three-thirds system' of land use is mainly responsible to guide the extent of agricultural land. In 1957 altogether 11.65 per cent of land area is cultivated which decreased to 11.23 in 1958. This change is noticeable due to "sow a smaller acreage, realize higher yield, and reap a greater harvest", launched in late 1958. Under the "three-thirds system", the communes were asked to use one-third of the existing farmland for growing principal crops, one-third to lie fallow or for growing fodder crops and green manure, and one-third to be dug into reservoirs or for growing trees and ornamental plants. However, several months later a policy to the opposite extreme was adopted, and the communes were urged to "sow a larger acreage and reap a greater harvest".

In 1960, 33 per cent of the cultivated land in China was low yielding, saline alkali, acidic, sandy, swampy or erosive. Low yielding cultivated land is found in Shanghai, Peking, Tientsin and the coastal area of northern Kiangsu.

China is inadequately forested. Thus in order to "make China a green country", the afforestation programme was launched along with the development of agriculture. The task of afforestation is given to the commune. In 1964 commune had more than 30,000 forest farms, with 2 lakh persons engaged in afforestation work. In this way agricultural producers' cooperatives are set task to plant trees in villages, and the State is responsible for planting timber forests, shelter belts for water and soil conservation, wind-breaks, sand-breaks and shelter belt along sea coasts. It is helpful in advancing agriculture as a public service to protect land against erosion.

TABLE 7.7 : **Cultivated land in China, 1949-58**

Year	% of Cultivated Land
1949	10.20
1950	10.45
1951	10.80
1952	11.24
1953	11.31
1954	11.39
1955	11.47
1956	11.64
1957	11.65
1958	11.23

Source **: State Statistical Bureau, Peking, 1960, p. 128.**

TABLE 7.8 : **Afforestated areas in China (1950-59)**

Year	Total forest cover in 000' hectares
1950	127
1951	453
1952	1091
1953	1118
1954	1172
1955	1719
1956	5752
1957	4377
1958	17547
1959	18760

Source : State Statistical Bureau, Peking, 1960, p. 133.

China's deserts total about 177 lakhs hectare, an area nearly equivalent to the total area of existing cultivated land, and are distributed in the Sinkiang-Vighur, Inner Mongolia, and Ningheria-Hui autonomous regions and in Chinghai, Kansu and Shenri provinces. The principal deserts include Taklamakan desert, the Kurbantongut desert in Sinkiang, and the Tengri desert, the Patachilian desert, and the Ulanpuho desert in Inner Mongolia. China's government believe that 3,350,000 hectares of the deserts can be improved to grow some plant and developed into livestock bases.

China's grassland cover nearly 30 per cent of the country's total land area and is mainly found in Greater Khingan mountains towards Tibetan Plateau. It constitutes a major portion of Sinkiang, Chinghai, Inner Mongolia, Ninghsia, and the Kantze and Ahpo districts in Szechuan. Flat grassland constitutes two-thirds of the total area of China's grassland, while the rest one-third consists mainly of hilly grassland. Most of the grasslands lie in arid zone where the yield is unstable and the rainfall is deficient.

In 1949-60 droughts and floods have affected 1072 lakh hectares of cultivated land in China which did not produce anything. The greatest havoc was created by drought which affected 67 lakh hectares of land. Except Tibet and Sinkiang all other provinces and autonomous regions suffered from drought in different times and with varying intensity. Hopei, Honan, Shantung, Shansi, Shensi, Inner Mongolia, Kansu, Szechuan, Yunnan, Kweichou, Kwangtung, Kwangri and Fukien were affected by droughts in both spring and summer. Twenty provinces and autonomous regions were invaded by typhoons and floods in 1960, most seriously in Liaoning, Kirin, Heilungkiang, Kwangtung, Fukien, Kiangsu and Shantung. Eastern Liaoning heavily suffered from floods in the hydrological record of the province.

LAND USE CLASSIFICATION IN INDIA

In India various organizations and the geography department of the Universities are now holding land utilization survey and mapping mainly on regional level. First of all National Atlas Organization's work should be considered, which is preparing maps on a scale of 1:1,000,000 and printed on 1:5,000,000 (1957) and Diagnostic Survey of the Damodar Valley region (1968) prepared maps in 1:500,000 and printed on 1:1,000,000, These surveys were conducted under the guidance of S.P. Chatterjee and the mapping units adopted are based on the scheme earlier followed by him in Howrah Land Utilization Survey and also the scheme recommended by the Commission on land use of International Geographical Union. The schemes worked out are listed below.[28]

TABLE 7.9 : **Index of land utilization map (plate 16) National Atlas Organisation, 1957**

1. (a) Dense forest,
   (b) Open forest.

2. Scrub, 3. Arable land, 4. Arable land with trees, 5. Plantation, 6. Pasture, 7. (a) Rocky waste, (b) Sandy waste, 8. Alpine grass and serub, 9. Glaciated Region.

TABLE 7.10 : **Land utilization units (Map No. 18) showing land use region diagnostic survey of Damodar Valley region**

(1) Field crops, (2) orchards including smaller water features and land of non-agricultural uses, (3) Dense forest, (4) Light forest, (5) Non-agricultural use of land—miner and quarries (working and abandoned), (6) Unproductive land, (7) (a) Water features—tanks, (b) Larger villages

with houses 300 and above, (c) Culturable waste-small, (8) Culturable waste—large, (9) Smaller villages, (10) Cities and towns and their characteristics, (a) Administrative—commercial and Industrial, (b) Residential, (c) Industrial, (d) Mining, (e) Mining and Industrial, (f) Agricultural and Administrative.

TABLE 7.11 : **Before 1951, the land use statistics collected by the ministry of agriculture, Government of India were arranged according to the following classification:**[29]

(1) Total Geographical Area, (2) Area under forest, (3) Not available for cultivation, (4) Current fallow, (5) Other uncultivated land, excluding current fallow, (6) Net area sown.

TABLE 7.12 : **Directorate of economics and statistics (Land Utilization units)**

(1) Forest, (2) Land not available for cultivation, (a) Land to non-agricultural use, (b) Barren and uncultivable land, (3) Other uncultivable and excluding fallow land, (a) Permanent and other grazing ground, (b) Land under miscellaneous tree, crops and groves, (4) Fallow land, (a) Culturable waste, (b) Fallow land other than current fallow, (c) Current fallow, (5) Cultivated land; (a) Net sown area, (b) Area sown more than once.

TABLE 7.13 : **Land utilization mapping units proposed (Sen) for aerial photo-interpretation (Desert, 1968-72)**

(1) Forest, (2) Settlement, (3) Roads and Railways, (4) Land not available for cultivation; (a) Sandy waste, (b) Rocky waste, (c) Gravel land, (5) Hills, (6) Sloping hill sides (7) Fallow lands, (8) Cultivated land.

TABLE 7.14 : **E. Ahmad had classified**[30] **the rural land use into following categories with detailed colour techniques for land use mapping**

(1) Triple cropped land—deep yellow with thick green slauting stripes, (2) Double cropped land—deep yellow with thin green slauting stripes, (3) Single cropped land—deep yellow, (4) Fallow—light yellow; number of years fallowing should be written as 1, 2, 3, etc. (5) Land under grove and orchards light green and red dots; name of the prominent trees should be mentioned, (6) Land under shrubs and grasses—light green with blue dots, name of the prominent grasses should be written, (7) Land permanently under water-blue, (8) Built up land (village

buildings, schools, temples, etc.)—red, (9) Land under communication e.g. railways, roads, and tracks—orange, (10) Barren lands i.e., snow, steep slope, ravines, osar, sand, flood plain, village proximity, threshing ground, grave yards—white; type of waste should be written, (11) Forest—dark green with Re. for reserved forest, Pr. for protected forest and W. for village waste.

TABLE 7.15 : **While classifying the land use of canal irrigated area in Patna district, R.N.P. Sinha[31] has the firm conviction that his classification is most suitable under Indian condition. His classification and colour schemes are as follows :**

(1) Single cropped land: (a) Aghani—Light yellow with small green dots, (b) Rabi—Light yellow, (2) Double cropped land: (a) Bhadai and Rabi—Deep yellow with small green dots and dashes, (b) Bhadi and Aghani—Deep yellow with small green dots, (c) Aghani and Rabi—Deep yellow, (3) Triple cropped land—Deep yellow with big green dots, (4) Current Fallow—Yellow stripes, (5) Culturable waste—Broken yellow stripes, (6) Waste land—White, (7) Forest and woodland—Deep green, (8) Village groves and orchards—Deep green with small brown dots, (9) Bushes and shrubs—light green with small brown dots, (10) Grasses—Light green, (11) Built-up Area—Red, (12) Water Bodies—Blue.

The above scheme of colouring has the advantage that the various land use types will be quite distinguishable and the scheme corresponds with that of the topographical sheets of India which is quite familiar to us.

GENERAL LAND USE: BIHAR

Land is classified into six classes according to use—net sown area, current and old fallows, barren and uncultivable waste and non-agricultural uses, permanent pastures and orchards, forests and others. Pattern of land use is determined by two sets of factors, (a) The physical factors like topography, climate and soils which broadly determine the capabilities of the land and (b) the human factors like the length of occupance, density of population, social and economic institutions, etc. which determine the extent to which the resources of the land are utilized. The classification of land in Bihar in 1970-71 is given below:

TABLE 7.16 : **Classification of land in Bihar in 1970-71**

Categories	Percentage	Area in thousand hectares
Net sown area	52.51	3217.68
Fallows	13.33	817.20
Permanent pastures & orchards	4.86	298.26
Barren and uncultiv-able waste	10.20	625.50
Non-agricultural uses	6.16	377.50
Forests	12.90	790.48
Bihar Total	100.00	6126.62

The following table shows the broad pattern of average land utilization in Bihar :

TABLE 7.17 : **Percentage of land utilization in Bihar (1970-71)**

District	Net Sown Area	Fallows	Permanent pastures and orchards	Barren and uncul-tivable waste	Non-agri-cul-tural uses	Forests
Patna	77.0	4.0	1.0	14.0	3.0	1.0
Gaya	58.0	12.0	3.0	9.0	5.0	12.0
Shahabad	68.0	3.0	1.0	9.0	4.0	15.0
Saran	71.0	10.0	3.0	11.0	5.0	X
Champaran	68.0	6.0	2.0	12.0	2.0	10.0
Muzaffarpur	74.0	6.0	2.0	16.0	2.0	X
Darbhanga	69.0	8.0	3.0	18.0	2.0	X
Monghyr	49.0	16.0	4.0	10.0	8.0	13.0
Bhagalpur	53.0	12.0	5.0	9.0	13.0	8.0
Saharsa	60.0	17.0	6.0	11.0	6.0	X
Purnea	66.0	14.0	4.0	11.0	5.0	X
Santal Pargana	40.0	23.0	12.0	8.0	7.0	10.0
Hazaribagh	20.0	17.0	8.0	5.0	10.0	40.0
Ranchi	40.0	21.0	5.0	5.0	6.0	23.0
Palamau	21.0	21.0	2.0	2.0	6.0	48.0
Dhanbad	28.0	21.0	10.0	20.0	12.0	9.0
Singhbhum	30.0	17.0	8.0	7.0	9.0	29.0
Total	52.51	13.33	4.86	10.20	6.16	12.90

Source : Calculated from data furnished by the Department of Statistics and Evaluation, Government of Bihar, Patna.

Taking the net sown area and fallow lands together, cultivation is found to be extended to the farthest limits. There are considerable amount of forests in Chota Nagpur Plateau and only a small area is left under orchards and pastures. This is but natural in the region which has a long history of occupance and has a dense population. Forests survive only in Chota Nagpur Plateau and in the Sub-Himalayan zone of extreme north-west in Champaran where the percentage rises to more than 10 per cent. All other districts are marked by flat alluvial plains, the land under forest is less than 0.5 per cent. Similarly the barren and unculturable waste lands generally constitute 5 per cent or less of the total area, except in Monghyr, Bhagalpur and districts of Chota Nagpur (10 to 11 per cent) because of rugged nature of terrain in the latter. So far as non-agricultural uses of land are concerned, built-up area, roads and water bodies are major constituents under this head and the picture is practically the same all over the region.

Orchards and pastures occupy 1 per cent or less of the land in South Ganga Plain and Chota Nagpur Plateau and not more than 2 to 3 per cent of total land in North Bihar. It is only in the basins of the river Kosi in Saharsa and Adhwara group of rivers in Muzaffarpur and Darbhanga districts, that the land under orchards and pastures reach more than 3 per cent owing to the extension of cultivation to the farthest limits, little cultivable waste remains; the maximum being in Saharsa, Monghyr, Bhagalpur and Chota Nagpur Plateau (3 to 4 per cent). In the heavily populated districts of Patna, Shahabad, Saran and Muzaffarpur the percentage is less than one.

Fallows include two types of land, namely, old fallow (for 2 to 5 years) and current fallows. These are cultivated lands which have been left fallow because of rainfall deficiency or economic reasons in one year or other. Their extent, therefore, varies according to rainfall conditions in different years. Fallows are highest in Chota Nagpur Plateau (more than 20%) and in Saharsa (17%), Gaya (12%), Purnea (14%) and Monghyr (16%) where the land is relatively poor. In Monghyr, Bhagalpur and the whole of Chota Nagpur region, the land of inferior quality, and the percentage of land under fallow is relatively high (10 to 16%). Saran has 10 per cent fallow lands, probably because of extensive sugarcane cultivation in which land has necessarily to be left fallow periodically. Fallows range between 5 to 9 per cent in other districts, in Patna where they are 4 per cent of the total area.

The net sown area is around 70 per cent or more in Tirhut, Kosi and Patna division. Here the pressure of population is so intense that nearly every available patch of land is cultivated. Extension of cultivation is often at the expense of mango orchards and pasture lands. The presence of hills and uplands accounts for a lower percentage of net sown area in South Ganga Plain. Palamau has the least net sown area (21%) while Saharsa and Purnea which are no longer affected by the Kosi show a slightly better position than South Bihar.

When we consider the land use pattern in the region as a whole the net sown area constitutes 52.51 per cent and fallows 31.33 per cent of the total land. Thus almost 66 per cent of the area is cultivated. Orchards and cultivable wastes account for another 5 per cent of the total land. Uncultivable land is about a fifth of the total area, which is shared almost equally by built-up area and water bodies on the one hand, and forests and barren lands on the other, while the former is distributed almost uniformly over the region, the latter is confined to the extremities in the north and the south respectively. The intensity of cultivation is highest in the Ganga riparian tracts of the region.

CASE STUDY OF LAND USE IN BARAKHURD AND JHANWAN, NALANDA DISTRICT, BIHAR

Barakhurd and Jhanwan are revenue villages situated just in the west side of Noorsarai Block of Nalanda district. It occupies an area of 2532 acres. The land utilisation of these villages is determined by the nature of land having marked local variations in slope, distance from the settlement to the field, security of crops, soil fertility, ensured irrigation and availability of water which have made a great difference in the intensity of cropping near the settlement site and on the outskirts of the village (Fig. 7.4). Some parts of the village contain higher ground, some contain lower ground while some are fertile and others are not so. At some places irrigation is necessary, while at the other, it is not needed, so the

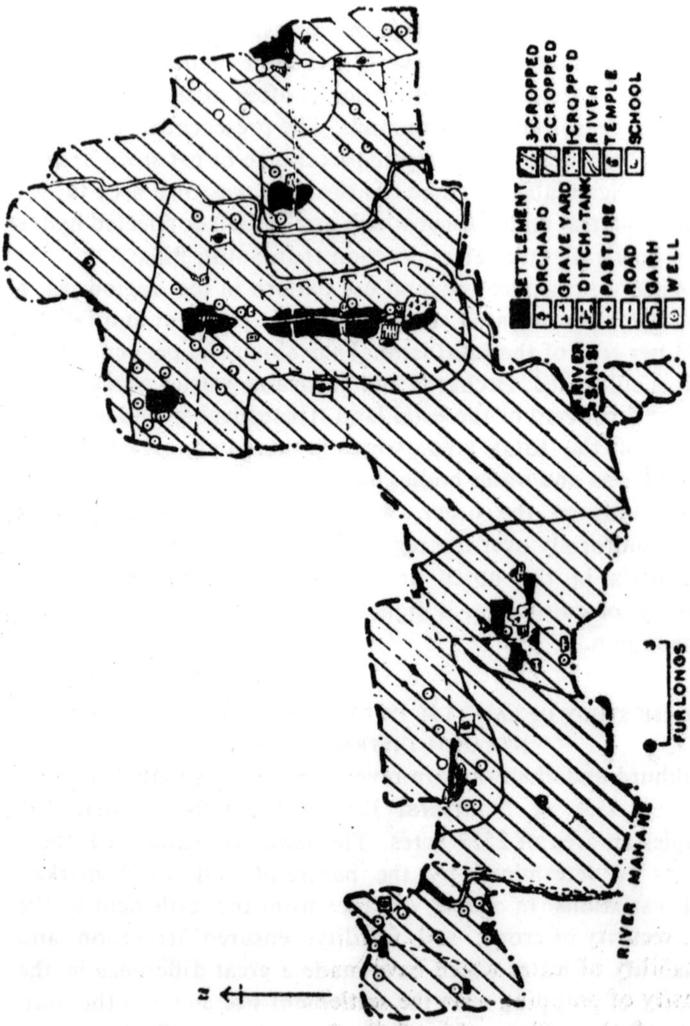

FIG. 7.4 : *Land Utilization in Barakhurd and Jhanwan*

different physical sets of land present diverse types of land use. Thus land use map not only shows the different use of the land but also indicates the path for future planning of the rural areas for the overall development of the cultural landscape.

Pattern of land use in Barakhurd and Jhanwan : Settlements occupy the higher ground of the area. It is divided into eleven groups or tolas viz. Barakhurd Bigha, Kharjama, Pandeychak, Nonia Bigha, Dharampur, Manera, Mahgupur Siwana, Malah Bigha, Jhanwan Dih and Jhanwan Bigha. All these are examples of semi-compact settlement type and all these tolas contain 110 acres of land from the total area of the village. These settlements are scattered on different sites with the availabili'y of drinking water. Each habitation followed by an ahar, suggests that man extracted the soil from that place for giving base to settlement site and plinth to the walls of houses.

Cultivable land : About 2100 acres of lands (82.95 per cent) is cultivable in these two villages. These lands are devoted to the paddy as Aghani crop, wheat, barely, gram, masuri and khesari as rabi crop, maize, millet and some early varieties of rice as Bhadai crop. Near the settlements on bari land ensured irrigation based potato and vegetables are the crops which are usually raised. The cultivation on such a large scale shows that by occupation most of the inhabitants are agriculturists and that is their basis of survival.

TABLE 7.18: **Barakhurd and Jhanwan land use pattern in 1971**

Land under different heads	Acreage under different use	Percentage of the total land
Settlements	110	4.34
Cultivable land	2100	82.95
Wasteland	76	3.01
Waterbodies	80	3.16
Orchards and groves	30	1.18
Pasture land	75	2.96
Current Fallow	61	2.40
Total Acreage	2532	100.00

Source: Furnished by the V.L.W. of Barakhurd and Jhanwan villages.

Single crop land : On the outskirts of the village, owing to greater distance and inaccessibility, lowland, less security of crops and lack of irrigation followed by irregularities of monsoon rain, not a single crop is raised on such type of land. Where the irrigation is not available, villagers harvest only paddy crop from fields and such fields are known as single crop lands. In recent years, by reason of the extension of irrigation and crash programme, the single cropped fields are changing fast into double cropped land.

Double crop land : Most of the land under cultivation occupied by double crop is related to irrigation or depends on the monsoonal rainfall. There are some fields which produce double crops as potato and paddy followed by potato II, gram and khesari in the next season. In this pattern, paddy is followed by masuri, wheat, barley and oilseeds.

Triple crop land : The fields near the village where well, surface-well and water from the pond is available, produce three crops of potato. These days people harvests Taichung Native I; Nagina and IR-8 paddies followed by the winter potato or oilseeds and then that field is left for garma-maize.

Multiple crop land : Multiple crop land is mainly devoted to vegetables, oilseeds and grama crops as interculture type farming which characterized by mixed cropping pattern. This pattern is available near the village site where small acreage of land has to fulfil every need of villagers.

Net sown area : In every season is highly fluctuating because the agriculture of this region is a gamble of monsoon and dependent upon the availability of water for irrigation. In these villages only small portion of land is left fallow in every season except summer when its prominence is quite apparent. This is due to the water-logging in low-land in rainy season and availability or non-availability of water in other seasons.

Waste land : The area above the pynes and both sides of river Mahane, Sansi and Itwan contain wasteland which are neither pasture land nor cultivable land. It contains nearly 76 acres (3.01%) of total land which are used as village roads leading to fields as inter-village connecting lines and embankments to the aforesaid smaller rivers.

Water bodies : Rivers Mahane, Sansi and Itwan are not perennial and they overflow their banks during rainy season. The amount of water is negligible because even a simple boat cannot float in February. The ponds, rivers, pynes, ahars and wells are the main sources of water for the agricultural operations. Each and every settlement consists of a water body so as to fulfil the villagers' need of water.

Orchards and groves : Forests are quite absent from these villages but orchards of mango are important which are followed by Sisso, bamboo and other types of useful mixed plantation. The higher ground above the pynes have mango and babul trees as in the south-east of Manera, the south of Dharampur, the west of Barakhurd, the east of Kharjama but the east of Siwana has the prominence of palm tree on the higher bank of river Itwan.

Pasture land : Pasture land is available on either side of the road connecting one settlement with another. These are used as grazing ground for animals but after harvesting of crops the field also serves the purpose of pasture land for some days until another crop is sown. Good pasture lands are not available as they contain small grass and are inferior in quality.

Fallow land : Old fallow is not available in these villages but current fallow varies from one season to another. The field restores its producing energy after fallowing. Its prominence is seen in summer only.

REFERENCES

1. J.K. Ableiter, Productivity rating in the soil survey report, *Soil Science Society Proceedings*, 16, pp. 416-422.
2. R.E. Storie, An Index for rating the agricultural value of soils, California, Agricultural Experiment Station, *Bulletin*, 556, 1933, Vol. 44. p. 3.
3. E.H. Graham, *Natural Principles of Land Use*, Oxford University Press, London, 1944, p. 97.
4. R.R. Renne, *Land Economics*, Harper & Bros, New York, 1947, pp. 35-36.
5. C.P. Barnes, *Land Classification Objectives and Requirements*, U.S. Resettlement Administration Land Utilization Division, Land Use Planning Publication 1, 1936, pp. 3-12.

6. P.E. James, *All Possible Worlds*, The Bobles-Merrill Company, INC, New York, 1972, p. 206.
7. P.E. James, *All Possible Worlds*, Chapter 13, p. 45.
8. L.D. Stamp, *Land of Tomorrow*, The Under Developed World, 1952, pp. 189-192.
9. A. Hillman, *Community Organization and Planning*, The Macmillan Co., New York, 1954, pp. 44-48.
10. F.J.A. Monkhouse, *Dictionary of Geography*, Edward Arnold, Bristal, 1970, p. 205.
11. L.D. Stamp, *Applied Geography*, Penguin Books, Suffolk, 1967, p. 48.
12. J. Kostrowicki, *The Typology of World Agriculture: A Preliminary Scheme*, Mimeograph, International Geographical Union, 1972.
13. J. Kostrowicki, *The Typology of World Agriculture: Principles Methods and Model Types*, Agricultural Typology and Land Utilization, Verona, Italy, 1975.
14. J. Kostrowicki, Agricultural Typology as a Tool in Planning Spatial Organization of Agricultural—IGU Commission on Agricultural Typology, *Geoforum*, Vol. 7, No. 3, 1976.
15. K.J. Beek, and J. Bennema, Land Evaluation for Agricultural Land Use Planning, an Ecological Methodology, Mimeograph, 1972.
16. J. Kostrowicki, op. cit., 1975.
17. Ibid, pp. 446-447.
18. L.D. Stamp, *The Land of Britain, its Use and Misuse*, 2nd Ed. Longmans Green, London, 1950, p. 34.
19. Alice Coleman, The second land use: Progress and prospect, *Geographical Journal*, vol. CXXVII, 1961, pp. 168-173.
20. T.W. Freeman, *Geography and Planning*, Hutchingson University Library, London, 1968, p. 63.
21. F.J.A. Monkhouse, *A Dictionary of Geography*, Edward Arnold, Bristol, 1970, p. 204.
22. L.D. Stamp, *Applied Geography*, Penguin Books, Suffolk, 1967, pp. 107-111.
23. L.D. Stamp, *Applied Geography*, Penguin Books: Suffolk, 1967, p. 105.
24. Kohnke and Bertrand, *Soil Conservation*, New York, 1959, pp. 152-154.
25. "Land Classification in the United States", National Resource Planning Board; U.S. Government Printing Office, Washington, D.C., March, 1941, pp. 1-3.
26. M.B. Nanavati, *Readings in Land Utilization*, The Indian Society of Agricultural Economics, Bombay, 1957, p. 96.
27. J.L. Buck, *Land Utilization in China*, University of Chicago Press, 1937, Chapter-VI.

28. A.K. Sen, Categories of land utilization units in arid zone, *The Deccan Geographer*, Vol. XII, No. 1, 1974, pp. 61-72.
29. V. Nath, Land utilization in India, *Journal of Soil and Water Conservation in India*, Vol. 1, No. 2, 1953, pp. 4-18.
30. E. Ahmad, *Geographical Essays on India*, Patna, 1954, pp. 15-16.
31. R.N.P. Sinha, Study of Land Use in a Canal Irrigated Area of Patna District, Unpublished Ph.D. Thesis, Patna University, 1959, pp. 177-79.

8

Role of Remote Sensing and Air-Photographs in Land Use Analysis

THE LAND presents a variegated picture with its huge mountains, remote deserts, and vast plains; its inland waters, forests and megalopolises. Surprisingly, in the present twentieth century, when the favourable conditions of explorations are enough in comparison with the past, many parts of this globe are poorly and inaccurately mapped. The exploding world population and the rising expectations of nearly all individuals in the face of dwindling proven reserves of natural resources, combine to make the creation and maintenance of truly modern maps which is really one of the most important tasks of this decade.

Today, the transmission of pictures and other geographical data to the ground by radio or television is quite common. Pictures may show anything from isolated hills and mesas to the whole mountain chains, from individual orchards to huge forests, glacier, from isolated hamlets to sprawling urban agglomerations. These, and innumerable other geographical features are readily by a variety of characteristics that can be sensed from above and transmitted to the ground control stations on command.

The satellite overview is leading to great improvements in map-making. Coverage from conventional aerial mapping is almost always discontinuous and the altitude limitations of the aircraft prevent synoptic views of the land below. In unaccessible areas the coverage is usually spread across seasonal boundaries and time intervals of decades. To construct a small scale map of India from aerial photographs would take at least several years, as five hundred thousands to one million photographs would be required for the photo mosaic.

A satellite in orbit, in contrast, can do the same job in a matter of weeks with four hundred to five hundred photographs. ERTS 1, for example, took thousands of cloud free photos each covering an area of more than sixteen thousand square kilometres.

At a meeting of the Fourth Annual Earth Resources Programme Review (January, 1972) at NASA's manned spacecraft centre at Houston, U.S.A. Gerlach predicted that up to 1977 it would be possible through ERT 1 photography to delineate land cover and land use for the whole world at a scale of 1:250,000. It could then be used "for national and regional planning, for monitoring and measuring trends, and hopefully for predicting and evaluating desirable modes of future land uses".[1]

In India first of all a small remote sensing project has been carried out to determine the coconut root disease. Aerial photographs of coconut plantations for a few localities in Kerala were taken from a helicopter wing, black and white and Ektachrome colour films with a 70 mm Hassebblad camera. The photo-interpretation has shown interesting reasons. It is important to note that some trees apparently healthy-looking but that recorded a weaker infra-red reflection on the film, were found to be infected by the virus.

The Indian Space Research Organization in Ahmedabad has an infra-red scanning camera which will be used for thermal mapping of oceans and land areas with the help of aircraft. The National Geophysical Research Institute in Hyderabad has been using air-borne magneto meters and electromagnetic sensors for mineral surveys on an experimental basis. The Geological Survey of India has been conducting aerial surveys using air-borne microwave transmitters for mineral surveys on an operational basis. The Indian Meteorological Department is studying the variations of snow-cover over the Himalayas with the help of satellite pictures on an experimental basis to estimate the contribution of snow melt to the North Indian rivers.

Remote sensing from aircraft and satellite platforms for the earth's resource is a new field, in which the advanced countries have a lead over India by 5 years. The methods of remote sensing survey is new and it is essential for the better appraisal of land use and natural resources of the Indian sub-continent.

AIR PHOTOGRAPHS AND LAND USE STUDIES

Today most of the Department of Environmental Sciences all over the world are actively engaged in the analysis of aerial photographs especially in geographical and economic studies. This is due to the rapid progress in spectro-photogrammetry.[1] In planning the region, resource appraisal, land use mapping, geology, geomorphology, forestry, pedology and agronomy, the students and research scholars have found it helpful in their research. In underdeveloped countries the hindrance is created by the non-availability of airphotographs to undertake such studies. In India Air-photo Interpretation centre, Dehradun, 'Air Force Agency, Calcutta' and several other agencies are providing air-photographs and training for survey and interpretation to the various institutions and the scholars of various disciplines.

Interpretation of Air-photo has the advantage of handling a large scale maps in lesser time with greater accuracy in depicting the various types of land use of wider space. It has also the advantage of depicting field boundary and for this fact it could be conveniently used in assessing the land tenure, providing base for preparing regional plans, master plan for urban areas and analysing the land use of an area. Hence air-photographs can be used both as map and data. Even a bird's eye view of such photographs by Stereoscope provides a lot of information regarding the landscape. In stereoscopic view the familiar scenes of the ground are transformed into lines and tones of varied faces where the true picture of landscape could be viewed. The technical advancement of panchromatic emulsions, use of polarized filters, infra-red and colour photography have made the techniques of photo-interpretation easier and provided a scientific base for land use analysis. Humanly such accuracy is impossible with the same team of investigation for months of costly field-work. In view of the need of land utilization maps through air-photographs, it is felt that the methods, which have been applied for Cyprus[2] in

London School of Economics would yield useful results. This study aims to describe the materials required for survey the stages by which the land use maps prepared and interpretation made.

Method of study : Air photographs are examined with the help of stereoscope. A lens stereoscope having a good magnification (X 4) is helpful in the identification of the types of vegetation settlement, etc. The transitional areas between different types of land use are generally drawn with the help of air photographs and are checked at doubtful and crucial point wherever confusion arises. A preliminary land use map is prepared in this manner. At this stage field check and ground studies are helpful when few sample checks are made for each of the different types of land use. The purpose is to correlate photographic characters of each type of land use with ground observations, and, to check complex areas of land utilization. In case of the completion of mapping all these boundaries are transferred with the help of a draughtsman to a 1:63,360 base. A detailed study of a particular type of land use can also be possible with the help of air photographs, which includes settlement types (rural or urban), house types, roads, lanes, footpaths, etc. The water-bodies which are used as fishing ground, depth of water (approximate) and character (perennial or seasonal) are generally determined. The doubtful issues are rechecked in the field. In this way, nearly 80 per cent information collected from air photographs, and the field study is reduced to a smaller one. Hence it is helpful in covering wider space in shortest possible time.

Specification of requirements : *Air photographs* : Photographs are generally taken in several strips. In doing so the turning, twisting, tilting and warping of flights are quite often distort the scale of the photographs. Such photographs should be vertical with at least 80 per cent fore and aft overlap providing stereoscopic cover of the whole area. The large scale photographs are better in quality and highly accurate in providing a huge amount of land use data. The air photographs of Cyprus available in London School of Economics ranges from R.F. 1:10,000 to 1:13,000. For this fact the print of any smaller scale would be of little use. In areas of simpler land use pattern such maps are useful e.g. in eastern coast of India where the cultivated land is mostly paddy, orchards of mango and coconut, etc. On the other hand, in Meghalaya where shifting cultivation is practised, interpretation have to be made minutely for patchy cultivation into bushes, and in such a situation photographs of larger scale i.e. 1:10,000 is necessary. Such air photographs should be clear having no hazy spots, shades of cloud and oblique cases.

Topographical maps : The land use data from air-photographs can be plotted on the topographic maps such as Cadastral sheets, on which field boundaries are clearly marked. In India the best map is 1:7920 anchal level which records settlements, roads, tracks and water bodies clearly, but no field pattern. In Meghalaya the problem is intensified as not larger than 1/63,360 map is available due to rugged terrain and forest cover.

Stereoscope : It is of various types among which the pocket stereoscope is handy for air photo interpretation in comparison with larger stereoscope with binocular attachment.

Stages in the preparation of land utilization map : *Plotting of Photographs* : The air photo should be placed on a table in runs of the aeroplane covering one side of the area under study to the other. This plotting should be made on anchal level maps. In case, the number of first and the last prints are arranged in each run than the subsequent insertion makes the process easy. In case the runs are more than one then only photographs of clear image should be selected for interpretation.

Interpretation of photographs : This requires a good grounding in air-photo interpretation technique and the type of landscape available. The principal task of interpreter is to establish relationship between actual landscape features with that of the photographs and the interpretation of photographs with that of the topographical sheets, in order to find out the error due to unawareness of the ground features.[3] In 1951 G.M. Howe published an account of his work on photographs of north-west Cardiganshire.[4] Due to easy access he has personally checked the critical areas of interpretation by the ground. Field work is not essential in case the interpreter is well acquainted with the ground. On the other hand, such studies must be supported by any other source of first hand information[5] as far as possible.

Taking the help of densitometres now it is possible to have statistical analyses of air photographs. The measurement of tonal variations by densitometres has reduced the chance of subjective lapses on a larger scale. To convert the land imagery into digital units generally perception type of automatic spot scanner is used. The simulation of this data gives correlation, regression for more than one variable including dispersion and concentration of agricultural activities.

The crops standing in the fields are interpreted with the help of textural contrast, tonal variations and differences in crop density through standard air photo inventory procedure. In the interpretation of air-photographs of agricultural fields for preparing land use maps the following sixteen standard codes are generally used (Table 8.1).

TABLE 8.1 : **Standard code for preparing land use maps through air photographs**

Code	Colour	Particulars of Land Use
0.1	Yellow	Cereals
0.2	Pink saffron	Weeds (other than fodder)
0.3	Medium saffron	Fibres and oil-seeds
0.4	Saffron	Legumes (beans)
0.5	Light green	Weeds used for fodder
0.5	Medium green	Pastures (temporary)
0.6	Light grey	Fallows
0.7	Green	Pastures (permanent), barn
0.8	Dark green	Moors, routes (productive)
0.9	Violet	Vine-yards
1.0	Red	Orchard
1.1	Orange	Market gardening
1.2	Dark red	Gardens
1.3	Pink red	Other crops (tobacco)
1.4	Dark green	Moor and fallows unproductive)
1.5	White	Buildings and courtyards
1.6	Green	Forest
1.7	Black	Uncultivable, rocks and waste lands
1.8	Blue	Water bodies

Source : After A.K. Tewari, *Geographical Review of India*, 1977.

In brief the whole process of aerial photo-interpretation for land use analysis may be divided into the following points:

Recording the land use data on topographic maps : Data for individual field cannot be collected from air photography, because:

(i) the filed boundaries do not appear on the 1/50,000 scale photographs; and

(ii) the fields are so small that on reduction to 1/250,000 two adjacent fields different land use could not be determined.

Preparation of land utilization map : In London School of Economics while preparing the land utilisation map from air-photographs, it was thought that the final map prepared on a scale of 1/250,000 would be most convenient for field survey. The traces reduced photographically, to the above scale, especially air-photographs of Cyprus. For shading purposes eleven land use categories have been evolved and combined together to record mixed categories. These maps are prepared in black and white in order to present simpler maps.

Analysis of land use pattern : The patterns of land use is commonly determined on medium scale maps (1:25,000) followed by adequate ground checking and rechecking. The areas of net sown, pasture, fallow, garden, road, railway, graveyard, forest, water bodies, settlement site, harvesting and thrashing ground, etc. are marked on the base map. On air photographs the agricultural crop areas are shown by dark patches while light grey for current fallows. In this way, this saves time and money of the investigator. The land use pattern areas engaged in various uses are generally determined by planimeter.

Identification of crops in different seasons : In practice it is difficult to identify the seasonal variations of crops with the help of 1:10,000 scale photographs and its superimposition on cadastral maps is highly useful for identification of crops. The density of crops and tonal variations, which are primarily determined by densitometers are useful aspects of photo interpretation. The National Institute of Geography, Paris uses the scale of density of crops is as follows.[6] O-white, 1-2:

Grey-very clear; 3-4: Grey—clear, 5-8: Medium grey, 7-8: dark grey, 9-10: Very dark grey, 11-12: Black.

The interpretation of various agricultural crops, crop rotation, land classification and determination of land capabilities are recently developed practices of photo-interpretation techniques developed in Zaire, Uganda, New Guinea, France, U.S.A. and U.S.S.R. With the help of this techniques France has recently started agriculture census and published complete set of coloured land use maps.

Measurement of crop productivity : The measurement of agricultural efficiency and yield rate are commonly impossible on air photographs but experts of agronomy can easily say about the health, growth rate, yield and other items of crop productivity with the help of infra-red false colour satellite imagery. This also facilitates the assessment of soil nutrients and moisture, duration of sunshine on particular land, plant mortality, immunity to pest and disease, quantity of input, types of seeds used and yield rate, etc. Infra-red radiation and heat pattern techniques are widely used in the interpretation of forest resources and land resource inventory.

Study of land use change : In the United States of America, Switzerland and Canada air photographs are extensively used to measure the land use change. The photographs of different time periods are valuable assets in order to measure the advance and retreat of net sown area, loss or forward marching of desert areas and sand dunes, etc.

Interpretation of input parameter : In the realm of agricultural geography and land utilization the assessment of damages caused by insects and pests, soil erosion, degeneration of soil nutrients use of a particular type of fertilizer, choragraphy content of a plant, extent of mixed farming, intensity of irrigation, etc. are easily measured with the help of infra-red air photographs. The measures of agricultural innovation and operational research in the field of agriculture are possible with the help of air photographs and satellite imageries.

Land use planning : Nowadays air-photographs are the vital source to present the total view of the environment towards the planner in a briefest possible manner under a stereoscopic vision. The sharp perception of the regional planner clearly visualizes the ensemble of the whole in the miniature likeness of landscape.[7] In India the village level planning is generally tried but due to lack of the minutest detail of land use data, the purposeful planning is endangered to a greater extent. Recently, it is hoped that the interpretation of satellite imageries and air-photographs will usher an era of meaningful planning for land resource analysis, judicious use of land and integrated rural development through better farm planning.[8]

CONCLUSION

Remote sensing and air photograph techniques are applicable to land utilization survey widely. The Indian Space Research Organization (ISRO) and the National Remote Sensing Agency (NRSA) are actively engaged in conducting experiments on the application of airborne remote sensing techniques in land resource inventory.

On micro-level the stereoscopic vision of aerial photographs has added a new dimension in the study of land utilization and depicting the features of landscape which are utilized by man. Air photographic study has the advantage of utilizing large scale maps with greater accuracy in lesser time.[9] In this study the techniques of air photo interpretation in land use analysis has been given due consideration for the students of environmental sciences.

REFERENCES

1. Ordway III, Frederick I., *Pictorial Guide to Planet Earth*, T.Y.C. Company, New York, 1975, p. 35.
2. R.R. Rawson, and K.R. Sealy, Air Photographs and land utilization maps, *Geography*, Vol. XLII, Part-I, No. 195, 1957.
3. R.R. Rawson, and S.H. Beaver, Aerial photography and geographical studies, *Geography*, Vol. XXXII, 1947, p 131.
4. G.M. Howe, A Note on the application of air photography to the Agricultural Geography of north-west cardiganshire, *Geography*, Vol. XXXVI, 1951, pp. 15-20.
5. R.R. Rawson, and K.R. Sealy, op. cit.
6. A.K. Tewari, Air photo and rural landuse analysis, *Geographical Review of India*, Vol. 39, No. 2, 1977, pp. 158-164.
7. Ibid.
8. N.P. Ayyar, Photogrammetry, Photo-Interpretation and the

Indian Geography, *Applied Geography*, (Ed. R.L. Singh), NGSI, Varanasi, 1968, pp. 20-27.

9. F. Walker, The geographical interpretation of air photography *Geography in the Tweneth Century*, (Ed. G. Taylor), Philosophical Library, New York, 1953, p. 25.

9

Land Utilization in Terms of Agricultural Typology

KOSTROWICKI[1] and Beek and Bennema[2] have classified the agricultural land in close connection with the social and ownership characteristics, organizational and technical characteristics, production and structural characteristics in order to formulate the agricultural typology of land utilization. Kostrowicki tries to devise World typology of agricultural land use after considering the pattern, process and colour schemes of World land use survey (Fig. 9.1). Beek and Bennema's classification is suitable to evaluate the land in relation to suitability of ecological condition for better utilization of the land in order to satisfy human needs.

The changing character of agriculture from one region to the other is known as agricultural typology. The agricultural typology should not be misunderstood with the formulation of agricultural regions. Both typology and regions are dynamic and systematic concepts. Hence typology is known as similarities between regional units in terms of agricultural characteristics. The association of such characteristics may occur quite often in time and space.[3] Various characteristics which quite often determine the agricultural typology are distributed in space in a stratified manner. The distribution of resulting typology does not form a contiguous area, but units of a certain type are quite often dispersed in nature. On the contrary, a region is a spatial homogeneity concept having some territorial limitations and difference between established places on a particular part of the earth surface characterised by an association of ecological variables. The agricultural typology is a hierarchical concept in which lower number of

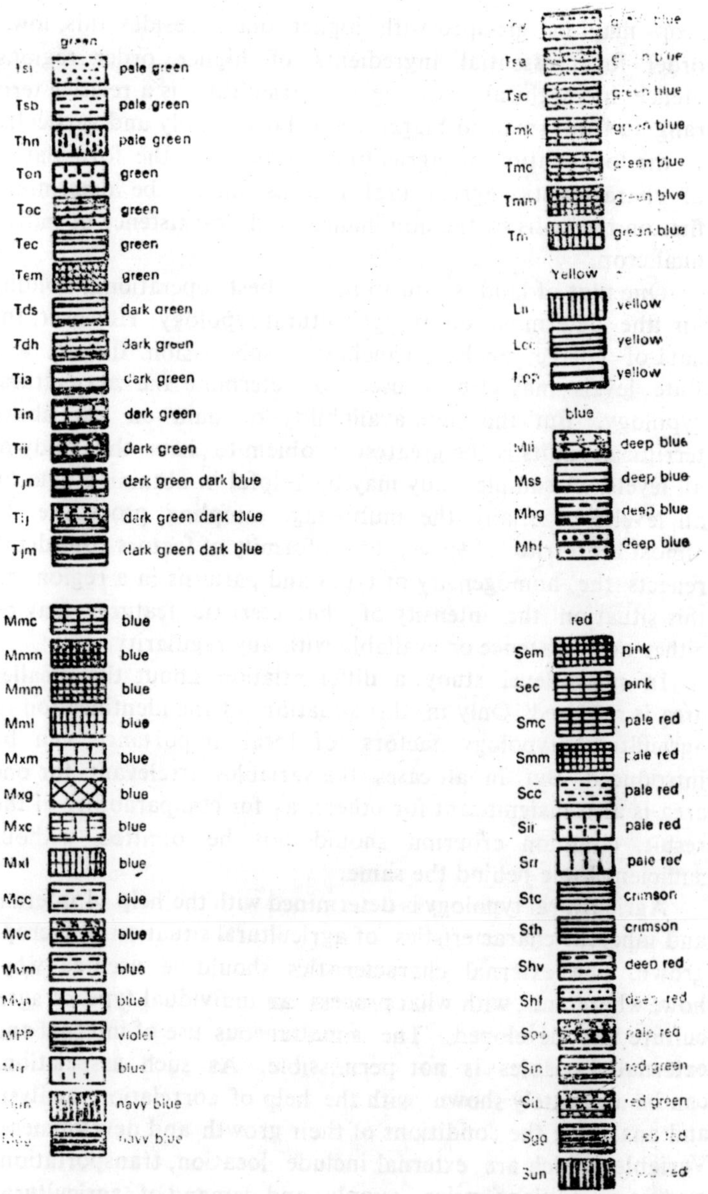

The legend proposed for mapping world types of agriculture comment: Colours—Principal four types of world agriculture: horizontal lines—crop emphasis; colour intensity—intensity of agriculture, asterisks—irrigation.

FIG. 9.1: *Kostrowicki's colour scheme of world agricultural typology*

crops may be grouped with higher one's, besides this, lower order form essential ingredients of higher order regions. Hence hierarchical ordering of agriculture is a relative term ranging from small to bigger both horizontally and vertically. In the formulation of agricultural typology, the formulation of characteristic agricultural regions should be determined first on the basis of the dominance and co-existence of individual crop.

One plot of land is said to be the best operational holding for the determination of agricultural typology. However, the data of village, anchal, panchayat, subdivision, district and state levels may also be used to determine the agricultural typology. But the non-availability of data on the above territorial units is the greatest problem to have the study of all levels. A sample study may by helpful in obtaining data on all levels, especially the multi-stage sampling procedure. In almost all territorial space, the uniformity of factors considered reflects the homogeneity of types and patterns in a region. In this situation the intensity of characteristic features may be either non-existence or available with any regularity.

In micro-level study, a differentiation about the smaller area is required. Only in this situation of the identification of agricultural typology factors of local importance can be introduced. But in all cases the variables irrelevant for one area is also insignificant for others, as for comparability of the result, common criterion should not be omitted without sufficient logic behind the same.

Agricultural typology is determined with the help of internal and inherent characteristics of agricultural situation and crops grown. The external characteristics should be used as why, how, where and with what process an individual type of agriculture has developed. The simultaneous use of internal and external variables is not permissible. As such associations can be accurately shown with the help of correlation analysis and analysing the conditions of their growth and development. Variables which are external include location, transportation, market conditions, price, supply and demand of agricultural products, but their individual condition change with the change of environmental conditions, time and space taken together. These factors will help in understanding the distribution of agriculture types only. Any change in the production of agricultural crops is followed by the change of external parameters essential for agricultural development. This also provides a basis for planning the spatial organization and rational types of agriculture. However, it is affected by the factors of input and output, organizational and technical characteristics and the use of fertilizer and irrigation facilities as well as these determine the amount of the production of food crops. The agricultural efficiency is determined by the fertility of the land, the amount of labour input, implements and seeds used and the system of land tenure. All these in most cases boost up agricultural production. Hence, their indentification is essential in ascertaining the types of agriculture.

The agricultural characteristics for determining typology can be grouped into the following :

(a) social and ownership characteristics,
(b) organizational and technical characteristics,
(c) production characteristics, and
(d) structural characteristics.

The first characteristic answers about the land owner and the scale of operation whether the same is small or large. The second, about the characteristics of labour and material input along with the operation of land holdings; and the third about the amount of production and the process of consumption; while the fourth one answers the proportion, combination of enterprise, livestock breeding, and gross agricultural production in terms of land utilization. The factors of social, land ownership, organizational, structural and input-output also help in the identification of agricultural typology. Actually speaking the agricultural statistical data are never found to be correct and they are insufficient to provide the material for agricultural typology. Hence the non-availability of data is a barrier in estimation of agricultural statistics for good results. The characteristics selected for agricultural typology should be expressed in quantitative terms. Although an investigator with better knowledge about an area may produce a good result of agricultural typology and the same may be reliable. But in most cases the same is impossible without proper agricultural statistics and methodology used. It is essential so that the comparisons can be made and the line of thinking may not be repeated for all the places and all the times. Statistical methods have the advantages of giving quick result instead of labour absorbing, traditional and slow moving devices.

In this way, the concept of agricultural typology should be treated as a highly interrelated phenomena or a complex system, so far as the system approach is concerned. The pattern, type and system of individual agriculture should be grouped according to similarities to get model types of agriculture. For these facts the concepts of agricultural system includes crop growing and livestock breeding for the production of food items. It also includes the classification of agriculture, land tenure systems, land use system, cropping system, systems of livestock breeding, farming, etc. In hierarchical concept it has been identified from individual agricultural holdings to World types of agriculture. As a dynamic concept agricultural typology changes in both evolutionary and revolutionary processes besides change in the basic attributes of agriculture.

METHOD OF STUDY

The method initiated by the IGU Commission at its fifth meeting on agricultural typology held in 1972 in Hamilton, Canada, recommended one method for agricultural typology studies irrespective of their scale, region and time. It has been clarified from a number of studies that (a) the study conducted with the help of variables as indicated by the commission does not affect the result much in any spatial dimensions and (b) the comparability of result is necessary in agricultural typology studies. Methods which do not guarantee such a comparability should not be recommended for use whether the same may be traditional, statistical, highly precise and multi-dimensional in nature.

Unmeasurable variables are denoted with the help of symbols for individual categories. On this basis, it is not possible to include them in the same category statistically. Hence the capital letter *O* denotes an absence of any typology. For this fact agricultural typology is represented after taking into account the following code:[4]

$$T = S \frac{O}{P} C$$

Where T is the type of agriculture,

S is the social and ownership characteristics and size of holdings,

O is organizational and technical characteristics,

P is production characteristics, and

C is structural characteristics.

Diagnostic features and variables: The commission of IGU on agricultural typology has selected 24 variables to measure the diagnostic features of agricultural types. The first two variables are represented by symbols, while the others are shown by indices, and kept their respective positions within the set characteristics. To determine the world agricultural typology five classes are considered in a uniform way which are as: 1—very low, 2—low, 3—medium, 4—high, 5—very high.

These twenty-two variables which represent the most significant agricultural characteristics were applied.

Social and ownership characteristics: (1) The proportion of agricultural land which is owned, controlled or held in common by a group of people under traditional customary tenure.

2. The proportion of agricultural land that is owned or held in owner like possession.

3. The proportion of agricultural land that is operated under the labour or share tenancy or any other form of landed bondage.

4. The rate of agricultural land operated by the consciously planned collective or state enterprises.

5. Size of holdings in terms of the number of actively employed per one holding.

6. Size of holdings in terms of the number of hectares of agricultural land per one holding.

7. Size of holdings in terms of the number of livestock measured in the number of conventional (large) animal units per one holding.

8. Size of holdings in terms of gross agricultural production per one holding.

9. Inputs of labour measured by the number of actively employed in agriculture per 100 hectares of agricultural land.

Organizational and technical characteristics : 10. Inputs of animal power measured by the number of draught animals in conventional (horse) units per 100 hectares of cultivated land.

11. Inputs of mechanical power measured by the number of tractors and self-propelling machinery in the number of HP per 100 hectares of cultivated land.

12. Chemical fertilizers measured as the amount of chemical fertilizers in pure content NPK used per one hectare of cultivated land.

13. Irrigation—measured as the percentage ratio of irrigated total cultivated land.

14. Intensity of crop land use expressed as the ratio of harvested to cultivable land (including fallow).

15. Importance of the system of perennial cropping measured by the percentage ratio of perennial cropland to total agricultural land.

16. Importance of the perennial grass land system measured by the percentage share of permanent grassland to total agricultural land.

17. Intensity of livestock breeding measured as the number of farm animals in conventional (large) animal units per 100 hectares of agricultural land.

Production characteristics : 18. Land productivity measured in gross agricultural (crop and animal) output in conventional (grain) units per one hectare of agricultural land.

19. Labour productivity measured as above per one actively employed in agriculture.

20. Degree of commercialization expressed as the percentage ratio of commercial to gross agricultural production.

21. Level of commercial production in commercial agricultural production in the same units as above per one hectare of agricultural land.

22. Degree of specialization measured by the method described by Kostrowicki,

Structural characteristics : 23. Orientation in gross production measured by the percentage share of livestock products in gross agricultural production.

24. Orientation in commercial production measured in the percentage rate of livestock products in commercial production.

TABLE 9.1 : Land tenure classification as measurable and non-measurable variables

1. Land ownership (Variable No.1):

Land is owned	Symbols
As none-measurable variable in common	
(by a tribe, class, village community, etc.) under customary rights	C
by a chief, feudal landlord and allied under customary rights	F
by a religious or other non-profit organization	R
by a private owner	P
by a corporate (business) organization	U
by a cooperative organization	V
by a collective organization	K
by a state (government, local authorities etc) organization	G

As measurable variable

Class	Percentage rate	
1	below-20	Under traditional customary tenure, share tenancy, private or joint property.
2	21-40	
3	41-60	
4	61-80	
5	over-80	

2. Land operation (Variable No. 2)

As none-measurable variable : *Land is operated*	Symbols
in common (by a tribe, class, village community members)	C
by an extended family	B
by slaves	S
by serfs under any form of landed bondage involving compulsory labour	F
by servile tenants (in exchange for services)	E
by sharecroppers	M
by tenants under fixed rent	T
by squatters	Q
by rent from tenants	N
by landowners	P
by part-time farmers	W
by hired workers	H
jointly by cooperative members	V
by members of a collective farm	K

As measurable variable

Class	Percentage rate	
1	below 20	Agricultural land operated by the consciously planned collective or State enterprises.
2	21-40	
3	41-60	
4	61-80	
5	over-80	

If there are two or more farmers of equal importance, then it should be treated as a mixed forming marked by two or more symbols (e.g. P/H or T/H).

TABLE 9.2: **Classes of measurable variables based on their world ranges**

Number of variables are the same as listed above

1	2	3	4	5
0-20	21-40	41-60	61-80	>80
0-20	21-40	41-60	61-80	>80
0-20	21-40	41-60	61-80	>80
0-20	21-40	41-60	61-80	>80
0-2	3-8	9-50	51-200	>200
0-5	6-20	21-50	51-200	>200
0-2	3-10	11-50	51-200	>200
0-100	101-900	901-10,000	10001-80000	>80000
0-3	4-15	16-40	41-150	>150
0-2	3-8	9-16	17-30	>30
0-6	7-15	16-36	37-90	>90
0-10	11-30	31-80	81-200	>200
0-10	11-25	26-58	51-80	>80
0-0,1	0,2-0,3	0,4-0,7	0,8-1,3	>1,3
0-10	11-20	21-40	41-60	>60
0-20	21-40	41-60	61-80	>80
0-10	11-30	31-80	81-180	>180
0-5	6-20	21-45	46-100	>100
0-40	41-100	101-250	251-800	>800
0-20	21-40	41-60	61-80	>80
0-3	4-12	13-30	31-80	>80
0-0,1	0,1-0,2	0,3-0,4	0,5-0,8	>0,8
0-20	21-40	41-60	61-80	>80
0-20	21-40	41-60	61-80	>80

Source: W. Tyszkiewicz, Types of agriculture in Poland as a sample of the typology of world agriculture, Reprinted from *Agricultural Typology and Land Utilization*, Verona, 1975, p. 394.

In areas of great variations from the proposed twenty-four variables for the determination of agricultural typology some additional variables of local character may be added as for example.[5]

—Percentage of hired workers to the total number of employed in agriculture,
—Percentage share of selected size groups of holdings in the total number of agricultural holding or in the total agricultural acreage,
—Degree of farm fragmentation,
—Total inputs of drought power in conventional units for unit area and the percentage of mechanical power in the total draught power,
—Organic manuring (expressed as the number of farm animals multiplied by conventional manual unit and by the percentage coefficient of manure used for agricultural purposes for 100 hectares of agricultural land),
—Total fertilization (chemical, organic, etc.) in conventional units per unit area and the percentage of chemical fertilizers in the total amount of fertilizers used,
—Percentage rate of drained land in the cultivated land,
—System of irrigation or water management system,
—Potential intensity of agriculture in conventional units,
—Land use system,
—System of land or crop rotation,
—Principal tools and machines applied in tillage and harvesting,
—Rate of land producing feeds to the number of animals in large animal units,
—System of livestock breeding,
—Final production per hectare of agricultural land and per one employed,
—Selected crop or animal products per one hectare of agricultural land or per one person employed in agriculture,
—Orientation in agricultural land use, livestock breeding and gross commercial production identified in a more accurate way either with one of the techniques that define leading elements in structure or with the following indices showing the percentage rates of:
—vegetative crops versus seed crops,

—Intensive crops in total arable land,
—Stucture (leguminous) crops in total arable land,
—Cattle in total number of livestock,
—Pigs in total number of livestock,
—Grain crops (in gross agricultural production),
—Fodder crops (in gross agricultural production),
—Vegetables in commercial production,
—Fruits in commercial production,
—Industrial crops in commercial production.
—Milk in commercial production,
—Food grains for commercial production.

It is quite apparent that with the increase of factors, the typology will automatically increase and that will also reduce the significance of other variables. In no case the investigator should be allowed to lessen the number of primary variables irrespective of their non-existence because that will reduce the chance of comparability for larger areas.

Model types of Indian agriculture : Agricultural typology on India has been determined after considering the various agricultural classification, case studies collected from the literature and field work done by Kostrowicki[6] and V.R. Singh. With the consideration of maximum class similarity in 24 agricultural variables, India can be grouped into eight categories e.g; Tsb, Tec, Tin, Tim, Tsc, Mii, Mhg, Mvm and Tsv in broader sense of the term.

In the scheme given below, four groups of characteristics have been grouped in the following way as the types of agriculture available in various parts of the Indian sub-continent.

(1) Tsb : Rotational bush fallow agriculture

$$1\text{-}2,\ 0,\ 2\text{-}3,\ 1,\ 1,\ 1,\ 1\text{-}2\frac{1\text{-}2,\ 1,\ 1,\ 1,\ 1,\ 2.\ 1\text{-}2,\ 1,\ 1}{1\text{-}2,\ 1,\ 1,\ 1,\ 1}\ 1,\ 1$$

It encompasses economically backward, culturally tribal and agriculturally poor states of India, such as Arunachal, Nagaland, Manipur, Mizoram, Tripura and Meghalaya. Jhuming cultivation is practised in all these areas but the effect of jute cultivation is found in Mizoram and Tripura.

In the above coded form, common land held by a chief through extended family of different sizes (2-3). Size of holdings (2) and less identifiable. Number of animals and per holding gross production very low (1). Inputs of labour per unit area—very low to low (1-2), inputs of capital very low (1). Duration of fallow shorter (2), share of perennial crops very low to low (1-2). Density of animal population very low (1), land productivity low (1-2), labour productivity and commercialization very low (1) and the dominance of crop products (1). The similar situation of agricultural typology is found in West Africa, South East Asia and Latin America.

(2) Tec: Current fallow, semi-subsistence crop agriculture

$$2441211 \frac{2111131}{2222221} 215111$$

This typology represents the areas of Jammu and Kashmir, Assam, Himachal Pradesh and Arunachal. In these areas the input of animal power, degree of specialization of permanent grassland and the level of commercialization and production orientation narrowly lag behind the normal value. This type of agricultural typology is found in some parts of Africa and Latin America.

(3) Tin: Labour Intensive, Semi-subsistence, Highly Productive, Non-Irrigated Agriculture

$$1241212 \frac{4311142}{4411112} 214111$$

This typology is found in densely populated States of Bihar, Orissa, West Bengal, Madhya Pradesh and Uttar Pradesh. This is also the case with south-east Asia.

(4) Tjm: Labour Intensive, Semi-Commercial, Irrigated and Medium Productive Agriculture

$$1151112 \frac{4422454}{3322332} 214112$$

This typology is specially found in Northern India.

(5) Tsc: Semi-commercial, Specialized Arid Land Crop Agriculture

$$1351221 \frac{3311232}{2211332} 324113$$

This is the clear case of agricultural typology found in arid lands of north western India, Pakistan and Sudan.

(6) Mii: Commercial Small Scale Non-specialized Irrigated

Highly Productive Oriented Towards Food Production Agriculture

$$1151211\frac{4234551}{5533444}114111$$

This sub-type is found in the States of Punjab and Haryana and outside India in Japan and Taiwan.

(7) Mhg: Commercial Small Scale Vegetable Growing Agriculture

$$1151222\frac{4145351}{5533555}115111$$

This type is predominantly observed in suburban areas of Bombay, Calcutta, Madras and Delhi etc.

(8) Mvm: Large Scale, Irrigated, Mixed Intensive Crop Growing Agriculture

$$1151244\frac{2154411}{4455554}113112$$

This typology is found in Andhra Pradesh, Karnataka, Maharashtra and Kerala.

(9) Tsv: Semi-Commercial Specialized Crop Agriculture

$$2241222\frac{3412241}{2222332}313113$$

It is observed in Gujarat and Tamil Nadu. In these states the production of cotton, sugarcane, groundnut and oilseeds are prevalent.

According to the assumption presented above, if a given case does not deviate from one of the model codes by more than ten variables, it should be considered as being of the same type of sub-type. In case the same deviates from the two or more model types by the same number of variables, such cases should be treated as intermediate type. In case a given model "deviates from all the model types by 11 or more variables, it should be considered either as being of a new type, not anticipated in the present scheme, or in case of aggreĝate units, it might result from a composite character of such a unit, and signify an average for two or more distinct types, which does not represent any one of those types. As such it should be investigated individually".[7]

As the model types analysed above do not exhaust all possible types of Indian agriculture, the cases with recurrent similar deviations from the model types can be described as new types of agriculture.

The typology discussed above do not claim to be static but a change is inevitable with the change of environmental conditions. The modification of ecological conditions may change the agricultural characteristics but with the inclusion of a new element the change in agricultural typology is impossible.

This study shows that the agricultural typology is the product of the association of elements available in a region. Kostrowicki suggested that it is a dynamic concept like the scene of a film. A change in typology is inevitable with the change of environmental conditions in larger dimensions but the introduction of one or two variable may not affect the result in general.

REFERENCES

1. J. Kostrowicki, The typology of world agriculture: A preliminary scheme, Mimeograph, International Geographical Union, 1972.
2. K.J. Beek, and J. Bennema, Land evaluation for agricultural land use: An Ecological Methodology, Mimeograph, 1972.
3. J. Kostrowicki, The typology of world agricultural principle, methods and model types, *Agricultural Typology and Land Utilization*, Verona, 1975, p. 431.
4. J. Kostrowicki, op. cit, 1975.
5. Ibid.
6. J. Kostrowicki, *World Types of Agriculture*, International Geographical Union, *Commission on Agricultural Typology*, Warsaw, 1975, pp. 24-35.
7. Ibid.

10

Methods of Land Capability Classification

INTRODUCTION

LAND capability classification is "a field investigation of soil properties, slope, degree of soil erosion and changing land use patterns which form the basis for future planning of soil and water conservation".[1] This study has been contemplated here to show the different degrees of productivity of the land and its suitability for raising different kinds of crops. The measurement of land capability gives a scientific judgment for the conservation of land. The conservation is also necessary to maintain its utmost capability and inherent productivity under particular ecological condition. Grouping of land on the basis of physical and cultural factors is indispensable for planning the agricultural economy through sound cropping system and conserving the land resource for its judicious future use. It also helps us to find out its efficiency for specified uses and proper diagnosis for a long time production. In this way, land use capability classification helps to organise soil factors for conservation[2] by considering the physical factors directly affecting the land productivity. There is a need for such studies in view of increasing population density and decreasing per capita supporting capacity of the land.

In the field of economic geography it is necessary to investigate the possibility of new methods of land capability classification on regional level for planned agricultural development. Most of the works done abroad such as: Northern Ireland system of land classification[3] land use capability classification of Iraq[4] and Cornell system of economic land classification,[5] land capability classification of the U.S. soil conservation service[6], agricultural land classification of the U.S.S.R.[7] are not were suited for measuring land capability in quantitative terms. Hence, a new method is designed here after measuring the quality of existing environmental condition and ecological balance of variables. The soil conservation has also been considered to focus the need and place for conservation practices to reduce production losses from erosion: to protect land from floods, to improve drainage[8] and to reclaim marshy grounds. Hence, this paper aims to devise a new method of land capability measurement to show its relationship with the net sown area and to suggest the land capability index required for a sound cropping system and intensive land use.

BIHAR: AS A CASE STUDY

Bihar has an area of 173876 sq. kilometre and 56 millions of total population as per the census 1971 which comes to a density of 324 persons per sq. kilometre. Geographically, this state is the land of productive riverine plain in the north and rugged terrain with forest clad mineralised belts of Chota Nagpur Plateau in the south. The Ganga divides the plain into two unequal parts of which the bigger northern plain is known as North Bihar plain and the smaller southern as South Bihar plain. These two plains are drained by numerous rivers like the Kosi, the Mahananda, the Burhi Gandak, the Kareh, the Kamla-Balan, the Chandan, the Chir, the Sone, the Poonpoon and several others. During the rainy season owing to overflow of these rivers, the entire Bihar plain comes under devastating flood every year. Once upon a time, grains. honey, milk and fish flowed like water from this region to the other regions of the country, but now with the increase in population and without the adequate exploitation of resources this has become the 'begging bowl' of India. Through this study an assessment has been made to measure the land capability of Bihar and suggest a rectification to weed out the hazards of the land in raising better crops and soil conservation. This is urgently needed to support the increasing population.

FACTORS AFFECTING LAND CAPABILITY

Land is the basic resource for mankind. To maintain its capability for producing more food crops to feed the increasing population, the study of land capability should be in the forefront of research in geography. Land is the product of nature. The capability of the land, thus, should be studied in close perspective of the physical elements of the environment, because human factors have transitory and lesser effects on land in increasing its productivity and capability. Owing to scientific innovations, the cultural influence on land is increasing fast, but everywhere in the third world economy physical factors predominate in determining the capability of the land.

To emphasize the importance of rainfall, it is quite often said that 'Indian agriculture is a gamble in the hands of monsoon' Hence, even in rainy season, there is no certainty about the amount of rainfall. The areas of highest rainfall lie in the north-eastern part of Bihar, especially in Saharsa, Purnea and Darbhanga districts. In these areas rainfall exceed 150 cms. per annum. The second zone is that of Chota Nagpur Plateau in the south which receives 130 to 149 cms. of rainfall per annum. The districts of Gaya, Patna, Sahabad and Saran receive a very low amount of rainfall below 119 cms. This amount is hardly conducive to late paddy and rabi crops.

Soil fertility comes next to rainfall. The Terai region of North Bihar has older alluvium, while the southern part has younger alluvium with a short strip of calcareous soil in between Saran and Champaran districts (Fig. 10.1). The South Bihar Plain has coarse texture and structure of alluvial soil, deposited by the rivers flowing from Chota Nagpur Plateau to the north up to the Ganga. Besides this, Chota Nagpur Plateau is wholly composed of forest and red soils, and is not favourable for intensive cultivation.

Irrigation facilities are poor in Bihar where only 22 per cent land is irrigated. The important means of irrigation are canals, tanks, tube-wells, wells and ditches, etc. Irrigation is provided, to some extent by the Kosi and the Gandak canal in North Bihar Plain, by the Sone canal in South Bihar Plain and by the D.V.C. Canal in Chota Nagpur Plateau. Tank irrigation is also prominent in plateau region and drier part of the western Bihar by collecting rain water. In North and South Bihar plains, tubewell and well irrigations have been quite common in recent years.

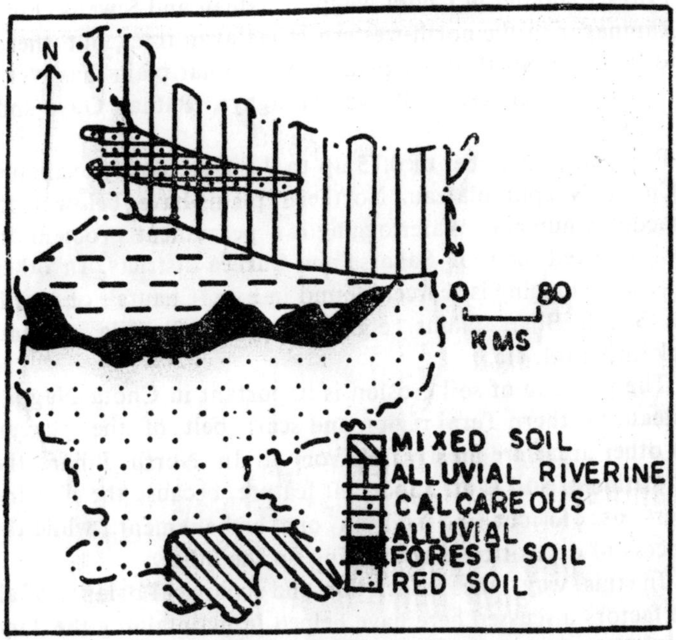

FIG. 10.1: *Bihar : Soil types*

Slope of the land plays a major role in determining the land capability, because the above 3 degrees of slope viz. in Ranchi, Palamau and Hazaribagh districts are lands of hills and valleys with only intervening patches of cultivated Don lands. With the decreasing slope towards Dhanbad and Singhbhum districts in the south-east, the intensity of cultivation increases. On the other hand, whole North and South Bihar plains have slope below 1 degree per kilometre, hence high land capability is the rule everywhere.

North and South Bihar plains are highly flood affected parts in the whole Ganga Basin, especially during rainy season. Most parts of Chota Nagpur Plateau and adjoining plains near northern scarp belt are flood free zones, while entire North Bihar is partially or wholly affected by the menace of flood. Just like flood, drought is the rule in Gaya, Rohtas, Palamau, Aurangabad and Nawadah districts of South Bihar plain.

Forests abound in Chota Nagpur plateau and Siwalik hills of Ramnagar in the north-western Himalayan tract, but there is no forest in North Bihar plain. South Bihar plain has very low forest cover especially in Monghyr, Patna, Gaya and Rohtas districts.

Ruggedness number of 0.15 up to 1.00 is mostly available in Chota Nagpur plateau. Northern plains have below 0.15 ruggedness number. Waterlogging is a permanent problem in Kosi ravaged belt of Saharsa and Purnea districts. In other areas waterlogging is scarcely found in big 'Chaurs' of North and South Bihar plains, e.g., Kabar tal, Baraila tal and Mokamesh tal, etc.

The problem of soil erosion is important in Chota Nagpur plateau, northern Terai region and scarp belt of the plateau but other areas are mostly free from it. In North Bihar, the deposition of soil is an important feature, because the flowing rivers use to cut the soil on one embankment, while the process of deposition is continuing on the other.

In this way, the association and ecological balance of all the factors discussed here have helped in maintaining the land capability of a particular region.

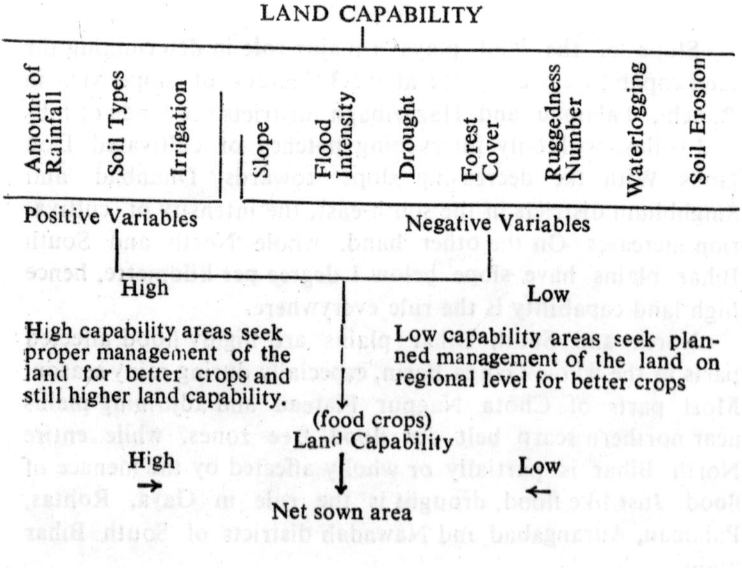

The above flow diagram shows that positive variables which help in producing higher amount of food crops as a measure of land capability, while negative variables are limiting factors of land use. Hence the areas which predominate in negative variables such as Chota Nagpur plateau produce less food crops in comparison with its areal extent. To increase the percentage of net sown area, the areas having predominance of positive variables like North Bihar Plains, seek proper nourishing of land so as to bring more land under cultivation to support high pressure of population. This is only possible with the provision of fertilizer, irrigation and deep tillage, etc. On the other hand, the areas characterized by negative land capability index seek planned management of the land on regional level for producing better crops. Such management includes the reclamation of marshy land, taming of the rivers, conservation of soil, forest clearance, reducing ruggedness of the land good tillage facilities and providing irrigation to the field easily in order to increase net sown area.

METHODOLOGY

Several methods have been commonly used for rating the land into capability classes.[10] Storie's[11] devised a land rating chart for California which assigns percentage values to the characteristics of soils and land features that affect land utilization and its productive capability. Marbut's (1935)[12] method is based on statistical information on yield history of the particular piece of land. Morgan's (1939)[13] considers the degree of adaptation of land types to various crops whereas Bennett's (1939)[14] method emphasises site and physical land factors. Stamp (1962)[15] has classified the land of Britain into ten major capability classes, on the basis of soil properties, land use and land values, while Choudhary (1968)[16] determined the land capability classes on the basis of soil properties and hazards in soil conservation. According to V.R. Singh, land capability classes should be determined with the help of physical and economic factors i.e., physiography, soil fertility. Power input, land use type, agricultural intensity and crop superiority. In this method[17] "the first step is to obtain, for each variable, data (D) as percentage (%) of the total area of each village or other unit area (Table 10.1). The percentage

TABLE 10.1 : Land use base

No. of village	Non-cultivable			Cultivable Waste			Arable Land			TR	RC
	D	%	R	D	%	R	D	%	R		
1	6.58	4.98	IV	23.75	16.02	III	69.66	21.79	III	10	3.33
2	4.62	3.50	V	10.98	7.40	VI	84.39	26.41	I	12	4.00
3	2.78	2.10	VI	15.65	10.55	IV	81.56	25.51	II	12	4.00
4	87.63	66.38	I	11.75	7.81	V	0.60	0.19	VI	12	4.00
5	18.95	14.35	II	60.42	40.94	I	20.62	6.45	V	8	2.66
6	11.51	8.69	III	25.65	17.30	II	62.82	19.65	IV	9	3.00
Grand Total	132.07	100		148.20	100		319.65	100			

Source **: V.R. Singh, Land capability: A method of classification, *The Professional Geographer*, 1970.**

(%) are then ranked (R) in descending order and rank coefficients (RC) are obtained by dividing the total of the rank values (TR) by the number of R observations. Rank coefficients (RC) of all variables are summed to give a composite index (CI). On the basis of the composite indices (CI), land well suited for agriculture can be distinguished from that not so suited. These two categories need further classification so that even the micro-regions representing homogeneous characteristics can be grouped together and possible combinations identified".

After due consideration of these methods of land capability classification the author has devised a new method after quantitative formulation of requisite variables.

The land has been classified according to its inherent fertility and productivity. To measure the land capability index, two sets of variables have been considered viz. (a) Positive e.g. soil fertility, amount of rainfall necessary for raising crops and better irrigation facilities and (b) negative variables which reduce the potentiality of land e.g. ruggedness number, soil erosion, water-logging, forest density, drought, higher degree slope and flood. Each variable is represented as a determinant of land capability. Most of the variables have been subdivided according to their position and importance in determining the capability of land either on positive and negative side. A score (25,15,10 and 5) has been allotted to each subclass of a variable separately on the basis of its importance for determining the positive and negative values of land in raising crops. After weighing properly (Table 10.2) these variables on regional level, the land capability index (Table 10.3) has been prepared and then the capability classes

$$LCI=\frac{\left(\sum_{i=1}^{n}\frac{R_1+S_2+1_3+\ldots\ldots+Q_n}{\Sigma PFV}\times 100\right)}{\left(\sum_{i=1}^{n}\frac{R_{n_1}+S_{e2}+W_3+\ldots\ldots+Q_n}{\Sigma NFV}\times 100\right)}$$

have been determined to see the regional condition. The formula on back page has been devised to measure the land capability index.

Where *LCI* is land capability index; *PFV* is positive factor values; *NFV* is negative factor value; *R*, *S*, *I* are rainfall, Soil fertility and irrigation facility respectively as positive variables and R_n, S_e, *W*, *F*, *D*, *H* and *T* are ruggedness number soil erosion waterlogging, forest density, drought, higher degree of slope and flood respectively as limiting factors of land capability.

TABLE 10.2 : **Weighing and evaluation of land capability**

Variable	Weighing values			
	25	15	10	5
Rainfall	Above 150 cm	149 to 130 cm	129 to 119 cm	Below 119 cm
Soil Types	Younger alluvium	Older alluvium	Sandy	Stony waste
Irrigation	Canals tubewells	Wells	Tank	Well & ditch
Slope	Above 3°	2° to 3°	1° to 2°	Below 1°
Flood	High flood	Partly flooded	Only in rainy season	No flood
Drought	Even water is scarce	Medium drought	Low drought	No drought
Forest Density	Above 20%	15% to 20%	5% to 15%	Below 5%
Ruggedness* Number	Above 1.00	0.15 to 1.00	0.05 to 0.15	Below 0.05
Waterlogging	Whole year	6 months	3 months	No waterlogging
Soil Erosion	High in plateau region	Medium in scarp region	Low in upland plain	Very low in plain land

Source : Compiled by the author.

* Ruggedness number (R_n) has been calculated by using the formula:

$$R_n = D_d \times \frac{RR}{5280}$$

Where D_d is drainage density and RR is the relative relief.

TABLE 10.3 : **Summary of land capability index in relation to net sown area in Bihar**

Regions/Districts	Positive Index	Negative Index	Final Percentage of L.C. Index	NSA to Total Area (1971)
Saran	66.66	34.28	1.94	73
East Champaran	45.66	37.20	1.25	53
West Champaran	53.33	54.28	0.98	45
Muzaffarpur	53.33	34.28	1.56	74
Darbhanga	59.99	34.28	1.56	57
Saharsa	73.33	51.42	1.42	40
Purnea	59.99	54.28	1.10	49
North Monghyr	59.99	40.00	1.49	50
South Monghyr	26.66	77.77	0.34	45
North Bhagalpur	73.33	42.45	1.70	64
South Bhagalpur	26.66	61.14	0.25	55
North Gaya	53.33	42.45	1.52	55
South Gaya	26.66	72.00	0.37	47
Bhojpur	66.66	34.28	1.44	70
Rohtas	46.66	77.77	0.60	45
Patna	80.00	54.28	1.47	78
Santal Parganas	20.00	88.57	0.22	40
Dhanbad	33.33	74.85	0.44	38
Hazaribagh	33.33	88.57	0.38	19
Palamau	20.00	88.57	0.22	10
Ranchi	33.33	88.57	0.38	38
Singhbhum	26.66	88.87	0.30	18

Note: NSA=Net Sown Area.

Source : Calculated by the author.

Table 10.2 shows the Relationship between land capability index in relation to net sown area. A product moment coefficient of correlation has also been estimated which comes to $r=0.67$. This alone signifies the relationship of the capability index and the net sown area available in the region. For these two components of land use the regression value comes to $y=0.22\times-0.10$ as shown in Fig. 10.2. In this way, it can be

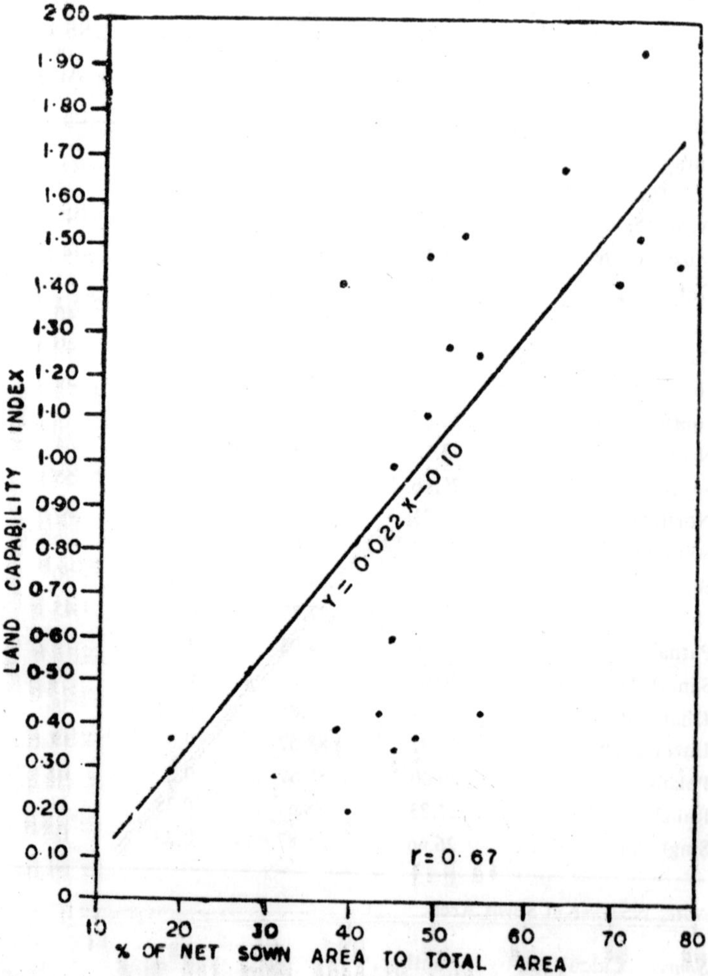

FIG. 10.2 : *Bihar : Relationship of land capability with net sown area 1971*

said that the intensity of net sown area is the real indicator of the potentialities of land capability in a region.

Afterwards, Lorenz Curve has been used (Fig. 10.3) to measure the dispersion of land capability index in different districts of Bihar. This graphical method has a merit to show the line of equal distribution of land capability in relation to the actual dispersion. For the preparation of Lorenz Curve the cumulative percentage of land capability and regions or districts has been given below (Table 10.4).

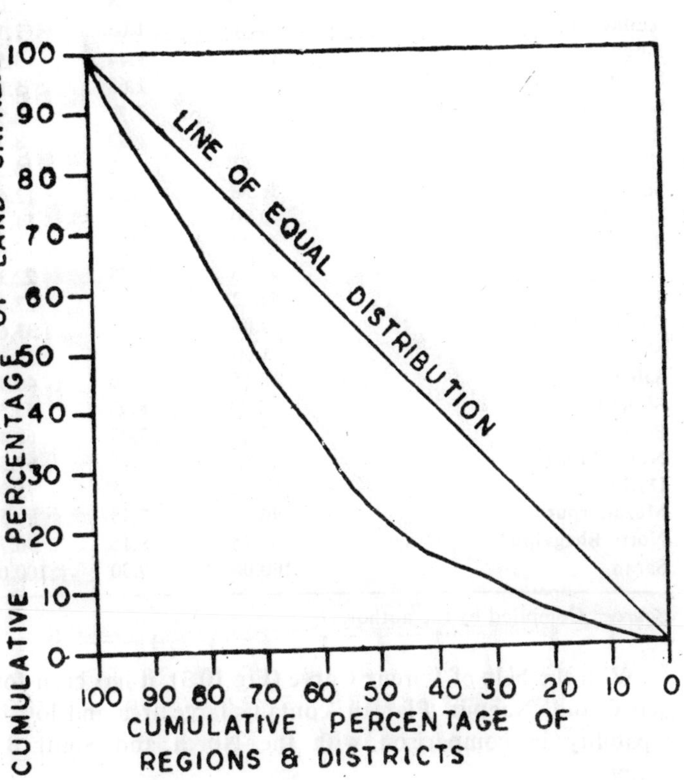

FIG. 10.3 : *Bihar: Dispersion of land capability on Lorenz Curve*

TABLE 10.4 : **Cumulative percentage of land capability and regions or districts in Bihar**

Regions/Districts	Cumulative percentage (x)	Percentage of land capability index	Cumulative percentage of land capability index (y)
Palamau	4.55	1.05	1.05
Santal	9.10	1.05	2.10
Singhbhum	13.64	1.44	3.54
South Monghyr	18.18	1.63	5.17
South Gaya	22.72	1.77	6.94
Ranchi	27.26	1.82	8.76
Hazaribagh	31.81	1.82	10.58
South Bhagalpur	36.36	2.06	12.64
Dhanbad	40.90	2.12	14.76
Rohtas	45.44	2.88	17.64
West Champaran	49.98	4.75	22.39
Purnea	54.52	5.27	27.66
East Champaran	59.06	5.99	33.65
North Gaya	63.60	5.99	39.64
Saharsa	68.15	6.80	46.45
Bhojpur	72.70	6.93	53.38
Patna	77.25	7.05	60.43
North Monghyr	81.80	7.15	67.53
Darbhanga	86.35	7.48	75.06
Muzaffarpur	90.90	7.48	82.56
North Bhagalpur	95.45	8.15	90.70
Saran	100.00	9.30	100.00

Source : Compiled by the author.

With the help of Lorenz Curve (Fig 10.3), it has been found that Chota Nagpur Plateau contains large area and low land capability in comparison with the North and South Bihar Plains.

LAND CAPABILITY REGIONS

On the basis of data as given in Table 1.2 a map of land capability regions (Fig. 10.4) in Bihar has been prepared. The categories of shading has been chosen on the basis of the 'Semi Inter Quartile Range' principle of statistics, which divides a set of data into four sub-groups. The calculation of the semi-inter quartile range requires two derived values: the value of land capability index (x) corresponding to the area under the distribution in which 25 per cent below the median (M) is known as quartile one ($Q1$), and the value of (x) corresponding to 25 per cent above the median is known as quartile three ($Q3$). In this way, the State of Bihar has been divided into four categories of land capability regions (Fig. 10.4).

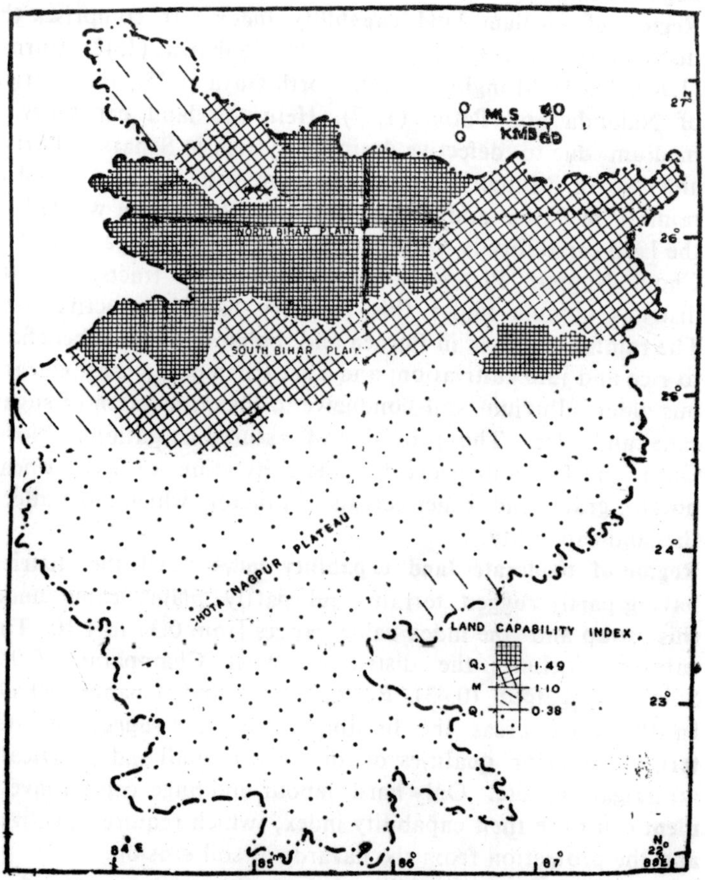

FIG. 10.4 : *Bihar: Land capability classification*

Region of highest land capability index : Districts having 1.49 land capability index (LCI) and above come under this category. It comprises the districts of Saran (1.94), Muzaffarpur (1.56), Darbhanga (1.75), North Bhagalpur (1.70), and Bhojpur (1.94). In these districts factors like fertile soil, considerable amount of rainfall for raising better crops supported by artificial irrigation practices keep the capability of the land highest and the positive factors predominate in comparison with the negative factors.

Region of medium land capability index : It comprises the districts of East Champaran (1.25), Saharsa (1.42) Purnea (1.10), North Monghyr (1.49), North Gaya (1.25), and districts of Nalanda and Patna (1.47). Here the land capability is medium due to defective drainage system in Saharsa, Purnea and North Monghyr and sandy loam with lower amount of rainfall in Patna and Gaya districts. But on the whole here the land capability is considerable better in comparison with Chota Nagpur Plateau and sub-Himalayan tracts. All the districts coming under this category have productive soils. The rainfall is high in Purnea and Saharsa which is beneficial to rice and jute cultivation; and East Champaran has calcareous older alluvium soil conducive to the production of sugarcane and rice. The provision of assured irrigation in North Gaya and Patna districts for the cultivation of rice, wheat, potato, gram and other cereals is a factor which most affects the land capability.

Region of moderate land capability index : All the districts having partly rugged terrain and partly plains come under this group and the index value ranges from 0.37 to 1.10. This category includes the district of West Champaran (0.98), South Bhagalpur (0.43), Rohtas (0.60) and Dhanbad (0.44). In the above areas the limiting factors are rugged forested terrain, inferior qualities of soils, low rainfall and practically no irrigation, etc. Only hard labour and huge capital investment can raise their capability index, which requires fertilizer, and the protection from the hazards of soil erosion.

Region of low land capability index : Stony waste and rugged forested terrain of Chota Nagpur Plateau come under this group. Here the capability value ranges from 0.22 to 0.38. The districts of Hazaribagh (0.38), Ranchi (0.38), Singhbhum

(0.30), Santal Parganas (0.22), South Gaya (0.37) and South Monghyr (0.34) come under this category. In these areas the negative factors which determine the land capability are high in comparison with the positive variables and for this reason these areas have lowest land capability index in Bihar.

LAND CAPABILITY INDEX REQUIRED FOR CERTAIN LAND USE

The cropping pattern and land use of a particular region can be determined by the capability classes, e.g., clay loam soil comes under class-one and will naturally be suited for crops like rice, jute, sugarcane, etc. while the loams and silt-loams are suitable for bhadai and rabi crops only. Under suitable water supply, sandy soils are also preferred for raising maize, sweet potato, root crops and vegetables. In southern part of Gaya and Rohtas districts under assured water supply of Sone Canal, summer paddy and hybrid maize are amply raised. On the sandy soil of annual inundation in the Gangetic riparian tract the root crops like water melon, etc. thrive well under unirrigated condition as moisture content on account of spill flood is correspondingly high.[18]

TABLE 10.5: **Land capability index required for certain land use, 1976**

Land capability required	Land use	Crops mainly raised
Above 1.50	Multiple cropping	Chillies, plantain, turmeric, coriander, cauliflower, cabbage and other vegetables.
1.00 to 1.49	Triple cropping	Sugarcane, wheat, rice, onion, etc.
0.50 to 1.00	Double cropping	Maize, millet, pulses, barley, jute, rice, groundnut, sweet potato, etc.
0.50 to 0.25	Single crop	Rice, kodo, china cotton, bajra, jowar, etc.
Below 0.25	Mostly barren and pasture land or forest cover	Rice in valleys and on mountain terraces, tea gardens, rubber, cocoa and coffee plantations on plateau region but only in rain-fed conditions.

Source : Compiled by the Author.

With the help of Table 10.5 one can easily determine the cropping pattern and the land use of a region or a field after perfectly measuring the positive and negative aspects of land potentiality. Thus, the land capability is determined by determine the potentiality of the land, the need for soil conservation and proper management of the land for planned agricultural development in a region.

At the end, it can be safely said that the term 'Land capability' is contemplated here as the scientific evaluation of the physical factors in enhancing the net sown area of Bihar. The product moment coefficient of correlation is high for land capability and net sown area, which suggests that land should be properly managed for boosting up the production from the land. Such measure is also necessary for farm planning on regional level to eradicate the regional imbalances of agricultural growth. The result obtained for particular land use, land capability required and cropping pattern sets out a very interesting picture. In this way, a new method of land capability measurement has been devised here in close relationship with net sown area, as a determinant. On Lorenz Curve the dispersion of land capability is also shown which represents the deviation from the line of equal distribution. This study has also a merit to help in agricultural planning and guiding for the better use of the land, in a particular capability class zone. Hence, it is expected that what has been given here may sound as an innovation in the field of land use, agricultural geography and planning.

REFERENCES

1. H.S. Sharma, *Land Capability Classification of the Lower Chambal Valley*, Symposium on Land Use in Developing Countries, 21st IGC, Aligarh, 1972, pp. 173-179.
2. J.H. Stallings, *Soil Conservation*, Prentice Hall Inc, Englewood Cliffs, N.J., 1957, p. 424.
3. Symons Leslie, *Agricultural Geography*, G. Bell & Sons Ltd., London, 1968, pp. 244-246.
4. W.L. Powers, Soil and land capabilities in Iraq, *Geographical Review*, Vol. LIV, No. 2, 1954, pp. 373-380.

5. H.E. Conklin, and B.F. Lucas, an economic classification of farm areas in Lewis country, New York, Cornell, Economic Land Classification, Leaflet No. 4.
6. S.P. Singh, *Dholri Village*, National Geographical Society of India, Varanasi, 1971, pp. 87-95.
7. G.I.P. Academician, The geographical study of agricultural land use, *Geographical Journal*, Vol. 124, 1958, pp. 452-463.
8. Land: The Year Book of Agriculture, 1958, *85th Congress*, No. 280, Washington, pp. 363-364.
9. R.E. Storie, An Index for rating the agricultural productivity of land, University of California, Agricultural Experiment Station, *Bulletin* 556, 1933, p. 3,
10. H.S. Sharma, op. cit., 1972.
11. R.E. Storie, *California Agriculture Experiment Station*, 1948.
12. C. Marbut, Soils of the United States, *Atlas of American Agriculture*, Part III, U.S.D. of Agriculture, Bureau of Agriculture, 1935.
13. M.F. Morgan, *Bulletin Connecticut Agriculture Experiment Station*, 423, 1939.
14. H.H. Bennett, *Soil Conservation*, McGraw Hill Book Company, INC, New York, 1939.
15. L.D. Stamp, *The Land of Britain: Its Use and Misuse*, 3rd Edition Longmans Ltd., 1962, pp. 362-83.
16. B. Choudhry, Determination of land capability classes and its application in the land use planning of the Subernrekha Basin, *Geographical Outlook*, Vol. V, 1968, pp. 95-102.
17. V.R. Singh, Land Capability: A Method of Classification, *The Professional Geographer*, Vol. XXII, No. 1, 1970, pp. 40-41.
18. K.N. Das, Agricultural landscape in the rural areas of Purnea and Saharsa Districts, Bihar, *Indian Journal of Regional Science*, Vol. V, No. 2, 1973, p. 113.

11

Agricultural Efficiency

"THE AGRICULTURAL production is the effects of farming".[1] In the field of geography, the concept of agricultural efficiency measurement is not a new aspect. The term agricultural efficiency connotes the productivity of a particular unit of land in foodgrains. It is a scientific device to study the inherent fertility, productivity and capability of the land, so that its misuse and underuse can be checked by planning for future use. The object of land use is to ensure the productivity and efficiency of agricultural land and it should be based on 'a real appreciation of agricultural needs, including the relative agricultural value of land in a locality'.[2]

Agriculture is the backbone of Indian economy, but owing to sedentary methods of cultivation and high pressure of population 'agriculture is not a business proposition for farmers, but living a way of life'.[3] There is a number of responsible factors which retard or put hindrance in increasing the agricultural efficiency. Among them misuse and underuse of agricultural lands, age old techniques of cultivation with outdated implements, keeping land away from chemicalization and rationalization, low living standards of peasants, defective tenure system and pest menace are altogether accountable for lower agricultural efficiency in Bihar.

It is quite often said that 'Indian agriculture is a gamble in the hands of monsoon'.[4] Thus, assured irrigation is necessary for good agricultural return. This is due to the vagaries of monsoon rainfall each and every year. The canal irrigation in the districts of Patna, Gaya, Shahabad, Eastern Kosi Canal area of Saharsa and Purnea and Gandak Canal schemes are praiseworthy attempts in this direction. The fragmentation of holding[5] is still another problem but the aim of land consolidation by the Government of Bihar up to 1978 is a good step to boost up our agricultural production. A significant fact is that Indian fields are not only small but are also scattered over a wide area.

During rainy season, the over-spill of water due to ice-melts in the Himalayan region and the heavy rainfall, North Bihar have to face the problem of flood every year. The government should immediately construct a number of reservoirs and mini-dams on 33 metres contour interval in the courses of rivers in Bihar to preserve the vast amount of water resource with tributary regulating canals for efficient use in dry season. Only through this way, the efficiency of our agriculture can be raised and we could save our land from devastating flood and enormous silting over entire North Bihar Plain.

Here the exact problem is how one should measure 'the relative agricultural value of land'. This can be solved as Bihar has large percentage (88%) of population engaged in producing cereals but a very small percentage supplements its daily food from sources other than cereals, e.g., egg, milk, vegetables, meat, fish and other eatables, which are locally available in rural areas. Hence efficiency of the land should be measured in terms of output per hectare.

METHODOLOGY

Several methods have been used to measure the agricultural efficiency with some virtues and shortcomings in them. Kendall (1939)[6] applied the method of ranking coefficient of yield of main crops per unit area for which the formula is as follows:

$$\frac{r_1+r_2+r_3 \ldots\ldots+r_n}{n}$$

Where r is the ranking of yield of individual crops and n is the number of crops. Previously this method was also applied in different countries of the world by Stamp (1960)[7] and on Uttar Pradesh by Shafi (1960).[8] In this method the ranking of crops is done simply with the help of per hectare production, without considering the total agricultural output.

Ganguli (1938)[9] devised an index of agricultural efficiency while measuring the agricultural efficiency of the Ganga Valley in India by multiplying the percentages of crops share with percentages of crop's yield in an areal unit and later averaging them in to one. It can be formulated into the following three stages:

$$lyn=\frac{y_1}{y}100\ldots\ldots(I)$$

$$Ein=(lyn.\ Cn)/100\ldots(II)$$

$$Ei=\frac{E_{i1}+E_{i2}+E_{i3}\ldots+E_{in}\ldots(III)}{n}$$

where lyn=%yield of crop n,

y_1=yield of individual crop in an areal unit;

y_r=yield of individual crop in the total area;

Ein=efficiency of crop n in an areal unit;

Cn=crop land share in %;

Ei=agricultural efficiency of all crops in an areal unit.

Later on Sapre and Deshpande (1964)[10] after modifying the Kendall's method an equation was suggested by multiplying ranking of crops with that of crop land share divided by the total of crop land share. The formula is as follows:

$$Ei=\frac{r_1c_1+r_2c_2+r_3c_3+\ldots\ldots+r_nc_n}{c_1+c_2+c_3\ldots\ldots+c_n}$$

where the notations are same as given above.

Bhatia (1967)[11] used the above efficiency index of Sapre and Deshpande with the some modification. He used Ganguli's percentage yield of crops (lyn) instead of yield ranking of individual crops (rn). His formula is as follows:

$$Ei=\frac{ly_1c_1+ly_2c_2+ly_3c_3+\ldots\ldots+ly_nc_n}{c_1+c_2+c_3\ldots\ldots+c_n}$$

The notations are same as given above.

M. Ali[12] used the method of S.S. Bhatia, while measuring the agricultural efficiency of Kashmir valley after considering the yield per hectare and proportionate share of land by an agricultural unit in comparison with the region as a whole.

In the same way Gupta, (1968)[13] while studying the agricultural efficiency of Cannanore district in South India presented a method for measuring the agricultural efficiency, after considering cereals and pulses, because they provide as much as two thirds of the Caloric intake of average Indian district. He measured the agricultural efficiency by considering yield efficiency of crops weighted by their proportionate share of land. The formula for measuring yield efficiency is as follows:

$$Er_1 = \frac{Yr_1}{YR} \times 100$$

where Er_1 = is yield efficiency,

Yr_1 = indicates the yield of crop r_1 per ha. in any unit area in calories, and

YR = is the yield of the same crop for the country as a whole in calories.

The yield efficiency of each crop is to be weighed by its proportionate share of land to the total area under all food crops in any areal unit. The agricultural efficiency in food production of a particular area may be measured with the help of the formula as follows:

$$Ia_1 = \frac{Yr_1 \times Cr_1 + Yr_2 \times Cr_2 + Yr_3 \times Cr_3 + \ldots + Yr_n \times Cr_n}{Cr_1 + Cr_2 + Cr_3 + \ldots + Cr_n}$$

where Ia_1 indicates agricultural efficiency of the area a_1; while Yr_1, Yr_2, Yr_3, etc. denote the yield efficiency of crops r_1, r_2, r_3 etc. and Cr_1, Cr_2, Cr_3 etc. indicate the proportionate share of land under crops r_1, r_2, r_3, etc. to total area under all food crops. These methods are praiseworthy attempts to measure the agricultural efficiency but the purpose of this paper is to give an alternative method of agricultural efficiency measurement, so that easily available data in season and crop report and simple calculation may open a new horizon with simple quantitative methods to study a difficult aspect of agricultural geography and land use planning.

Agriculture efficiency in Bihar has been assessed in this paper in terms of food grains[14] which comprise rice, wheat, barley, maize, ragi, jowar, bajra, gram, arhar, masoor and khesari. Per hectare production of all these crops in different districts and their summation value multiplied by 100 and divided by the total of all the districts in the region gives a value of location coefficient and that value serves here as agriculture efficiency. The formula is as follows:

$$AE = \frac{\Sigma D_1 (x_1 + x_2 + x_3 + \ldots\ldots + x_n \times 100}{\Sigma R (x_1 + x_2 + x_3 + \ldots\ldots x_n)}$$

where AE denotes agriculture efficiency, D_1 is district in particular, x_1, x_2, x_3 are crops grown in the region and ΣR is the regional total of all crops.

TABLE 11.1 : **Summary of agricultural efficiency in relation to net sown area in Bihar (1970-71)**

Districts	Total per hectare production in kg. for 14 crops	Location coefficient as *AE*	Net sown area in (1000) hectare	Location coefficient for net sown area
1	2	3	4	5
Patna	12107	7.31	278	8.64
Gaya	11718	7.07	211	6.55
Shahabad	10646	6.42	244	7.58
Saran	11010	6.64	256	7.95
Champaran	10694	6.45	243	7.56
Muzaffarpur	8441	5.09	265	8.12
Darbhanga	9216	5.56	251	7.79
Monghyr	8950	5.94	175	5.44
Bhagalpur	10183	6.14	192	5.96
Saharsa	7857	4.74	217	6.74
Purnea	9130	5.51	239	7.42
Santal Parganas	8979	5.42	145	4.55
Hazaribagh	8084	4.88	71	2.21
Ranchi	9072	5.47	142	4.41
Palamau	9408	5.66	77	2.45
Dhanbad	9428	5.68	103	3.28
Singhbum	9816	5.92	108	3.35
Bihar Total	165639	100.00	3218	100.00

Source: Season and Crop Report: 1970-71, Directorate of Statistical and Evaluation, Government of Bihar, 1972, pp. 96-99.

In Table 11.1 in addition to the calculation of location coefficient of food grains in different districts as agricultural efficiency index, the location coefficient of net sown area has also been determined. On the basis of these two data a graph (Fig. 11.1) has been prepared which shows the variation of agricultural efficiency over net sown area. Product moment correlation coefficient has also been determined, where r is 0.68 and the linear regression value is $Y=4.71+0.20x$. This relationship shows that agricultural efficiency has close relationship with the distribution of net sown area in Bihar as a whole.

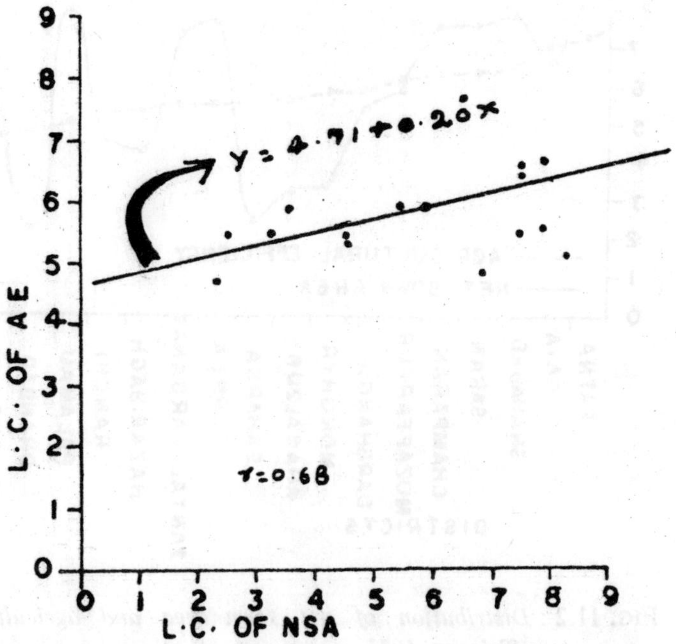

FIG. 11.1 : *Bihar: Relationship between net sown area and agricultural efficiency*

Similarly, an another graph (Fig. 11.2) has been prepared in which x axis is represented by the name of districts and on y axis the regional distribution of net sown area has been plotted along with agricultural efficiency. This gives a very interesting picture that in the districts of northern plains, the regional percentage of net sown area is high in comparison with plateau regions. On the other hand in plateau region agricultural efficiency is high in comparison with net sown area. This seems to be unusual that forested rugged terrain has higher percentage of agricultural efficiency in comparison with the low percentage of net sown area. The reality is that in Chota Nagpur plateau most of the cultivated lands produce paddy as single crop and its per hectare production is two to three times higher than barley, maize, marua and wheat mostly produced in northern plains.

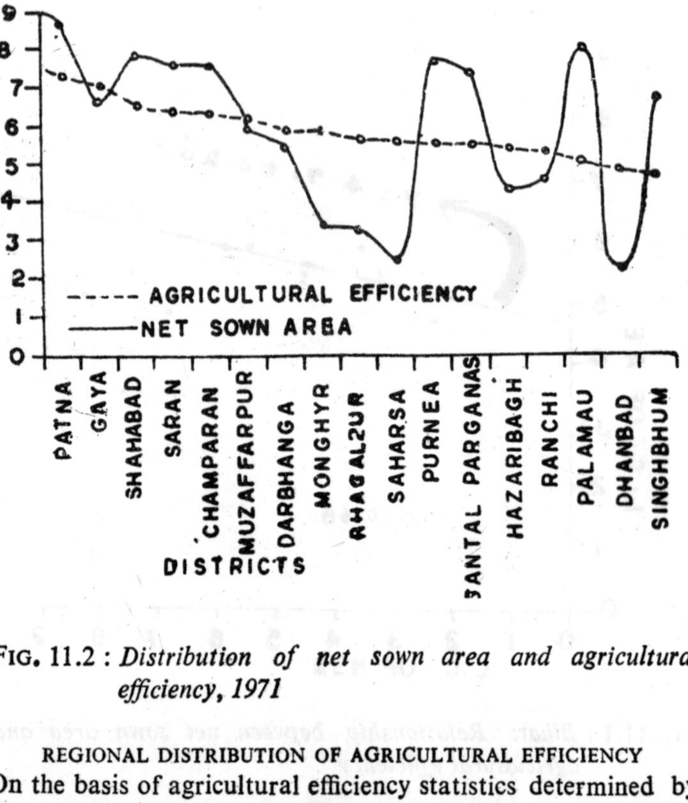

FIG. 11.2 : *Distribution of net sown area and agricultural efficiency, 1971*

REGIONAL DISTRIBUTION OF AGRICULTURAL EFFICIENCY

On the basis of agricultural efficiency statistics determined by the above mentioned formula are given in Table 11.1 a map (Fig. 11.3) has been prepared and shading has been chosen by the ranking of agricultural efficiency data, as the principle of Median and Quartile range, in which after determination of Median and Quartile-1 and Quartile-3 any set of data can be divided into four parts. Thus four grades of efficiency regions have been determined.

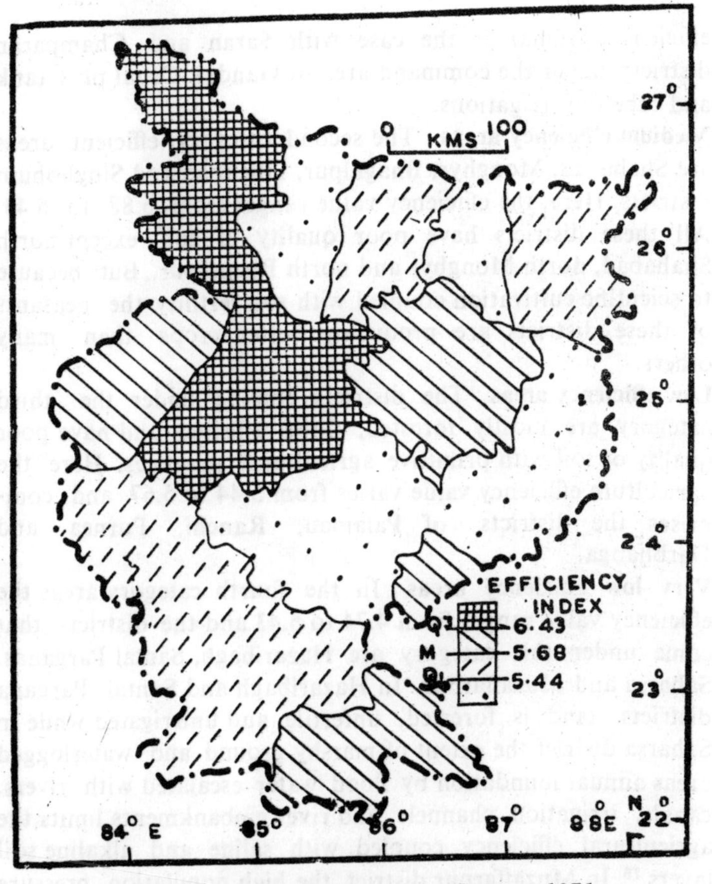

FIG. 11.3 : *Bihar: Agricultural efficiency 1971*

High efficiency areas : The first category area comprises the districts of Patna, Gaya, Saran and Champaran where the efficiency value ranges from 6.43 to 7.31. In the above mentioned fourteen crops per hectare production is higher than others. In Patna district owing to assured irrigation by Sone Canal and State tubewells of the government, private tubewells and with the excessive use of chemical fertilizers, the farmers in general are getting better production than others. In Gaya district, the network of Sone Canals, supported by private surface wells have sufficiently raised agricultural efficiency. Similar is the case with Saran and Champaran districts, under the command area of Gandak Canal plus tank and tubewell irrigations.

Medium efficiency areas : The second category efficient areas are Shahabad, Monghyr, Bhagalpur, Dhanbad and Singhbhum districts. Here, the efficiency value ranges from 5.88 to 6.43. All these districts have poor quality of soil except north Shahabad, north Monghyr and north Bhagalpur. But because to scientific cultivation coupled with soil fertility the peasants of these districts are producing better crops than many others.

Low efficiency areas : The districts coming under the third category are mostly forested, flood affected and have poor quality of soil with primitive agricultural practices. Here the agriculture efficiency value varies from 5.44 to 5.67 and comprises the districts of Palamau, Ranchi, Purnea and Darbhanga.

Very low efficiency areas : In the fourth category areas the efficiency value ranges from 4.74 to 5.43 and the districts that come under this category are Hazaribagh, Santal Parganas, Saharsa and Muzaffarpur. In Hazaribagh and Santal Pargana districts, land is forested, unfertile and unirrigated while in Saharsa district the extent of marshy ground and waterlogged areas annual inundation by flood water escalated with rivers, canals, irrigation channels and river embankments limits the agricultural efficiency coupled with saline and alkaline soil layers.[15] In Muzaffarpur district, the high population pressure and long continued agricultural practices since prehistoric time, have been exhausting the fertility of soil and so the agricultural efficiency is lowest there.

PROGRAMMES OF AGRICULTURAL DEVELOPMENT LAND USE

Agriculture occupies the most pivotal point in India's economy and the utilization of land. The aim of this section is to analyse programmes of raising agricultural efficiency and to suggest a suitable approach to make these programmes successful.

This programme is the outcome of raising agricultural production with the help of institutional agencies, relating education to agriculture science and providing incentive to the farmers. This includes[16] (1) the Community Development Programme, (2) Intensive Agricultural District Programme, (3) Intensive Agricultural Area Programme, (4) High Yielding Varieties Programme, (5) Multiple Cropping Programme (6) Small Farmers Development Agency and (7) Marginal Farmers and Agricultural Labourers Development Agency.

Community development programme : The government has considered the improvement of rural land use with the help of Panchayat Raj, cooperatives and agricultural development. The block headquarters were selected as 'nerve-centre' to carry the programmes of development to rural areas. The result of this programme is not satisfactory, although the underlying principles are sound. The causes of failure are improper location of block headquarter and curtailment in the diffusion of information in rural areas. Thus the block headquarters serve just the purpose of a small service centre where the quantity and quality of services provided are poor.

Intensive agricultural district programme : It is also known as package programme during the Third Five Year Plan on the suggestion of the Ford Foundation team. The team has given suggestion that intensive efforts for production should be made with a combination of all technological improvements and the supply of inputs in selected areas which are more favourable to boosting up production. This has been started in 1960-61 to induce the cultivator to adopt a package of improved agricultural practices with the use of improved seeds, fertilizers, pesticides, scientifically improved implements, proper soil and water managements, etc. so that the yield of crops may be raised. In this regard field demonstration is of considerable importance as it became the effective tool in motivating cultivators to adopt modern agricultural practices. Only a limited number of districts came under the purview of this programme in India.

Intensive agricultural area programme : This programme was launched covering the two-fifths districts of India. Owing to poor provision of fund and high coverage, this scheme yielded no significant result.[17]

High yielding varieties programme : The introduction of high yielding varieties was taken up as a full-fledged programme from the kharif season of 1966-67[18] in the selected areas of the country, where there was assured irrigation or rainfall, proper drainage, existence of strong input supply agencies e.g., cooperative societies, rural banks and several other voluntary organisations.

With the introduction of high yielding varieties "green revolution" has been achieved in the country especially in the production of wheat and rice. The evaluation studies of high yielding varieties have shown that nearly 90 per cent of the fields of wheat and rice under this scheme were benefited by the use of fertilizers, irrigation and improved implements in comparison with other crops. It has, thus, started a new strategy, a breakthrough in agricultural development which has contributed much to the intensive use of agricultural land to meet the demand of food grains.

Multiple cropping programme : Considering the increased use of irrigation, fertilizer and dwarf varieties of seed, the multiple cropping programme was taken up in India in 1967-68. The sole aim of this programme was to produce more per unit of area in a limited time period. The areas producing single crop in a year were largely benefited under this scheme after raising 3 and 4 crops in a year to fully utilize the farm inputs and to engage the labourers throughout the year. Especially the smaller farmers were benefited much under multiple cropping programme after hard labour, the full utilization of inputs besides increasing production per unit of area.

Small farmers development agency : The beneficieries of the above programmes are mostly affluent farmers neglecting the marginal ones. In spite of hard labour, the lack of fund is the main hurdle for poor peasants, for it deprives them of the use of the modern implements. On the suggestion of the All India rural credit review committee, the Government of India established this agency in 1970-71. Its prime aim is to guide, assist and coordinate the existing agencies work in order to help the poor farmers in a better way for the efficient and intensive use of the land.

Marginal farmers and agricultural labourers development agency : These two categories of labourers constitute the largest section of the society. This agency was set up in selected districts by the Government of India in 1970-71 to assist marginal cultivators in optimum utilization of their small land holdings by introducing vegetable growing, horticulture, dairy farming, poultry farming, etc. The main aim is to boost up production after channelizing agricultural credit loans, provision of fund, improved practices of agriculture and maximum utilization of land resources.

Thus after considering the above programme of agricultural development in India, the following criteria[19] may be used for measuring the agricultural efficiency:

Net farm income per acre : Gross returns minus costs of seeds, manures, fertilizers, fuel, oil, irrigation charges, repairs and depreciation, insecticides, hired bullock and human labour, hiring of machines, purchase of seed and land revenue should be evaluated for measuring farm efficiency. This measure indicates the returns to farm family's labour, capital including land and management. This measure is appropriate for comparing the efficiency of family based subsistence farms.

Family labour income per day : This should be estimated by deducting interest on the value of fixed capital from Net Farm Income. This measure indicates the income accruing to the farm family in lieu of providing labour and management to the farm.

Return to capital investment : This could be estimated by deducting value of family labour from the Net Farm Income and expressing in percentage terms. This measure provides a basis for comparing opportunity cost of farming.

Profits per acre : This may be computed by deducting the value of family labour and the interest on capital investment from the net farm income. This measure indicates returns to management.

Output-Input ratio : This may be computed by dividing gross returns by total costs.

The variation in agriculture efficiency in different districts is mainly due to various topographical features, fertility of soil intensity and amount of rainfall, irrigation facilities and development of transport routes in Bihar. Generally the northern plains are producing better crops than the plateau regions of Chota Nagpur, because the former have fertile soil in comparison with the later. In these areas the programmes of agricultural development act as catalyst in increasing the agricultural efficiency of a region.

REFERENCES

1. J. Kostrowicki, Geographical typology of agriculture in Poland, methods and problems, *Geographia Polonica*, Vol. 1, 1964, p. 127.
2. J.N. Jackson, Surveys for town and country planning, Hutchinson University Library, London, 1963, p. 99.
3. M.B. Nanavati, and J.J. Anjaria, *The Indian Rural Problem*, Vora & Co. Publishers Ltd., Bombay, 1951, p. 7.
4. R.B. Mandal, *The Geography of Rural Settlements in Bihar Plain*, Unpublished Ph.D. Thesis, Patna University, 1972, p. 46.
5. M.K. Elahi, Efficiency of agriculture in West Pakistan, *Pakistan Geographical Review*, Vol. 20, No. 2, 1965, pp. 66-116.
6. M.G. Kendall, The geographical distribution of crop productivity in England, *Journal of Royal Statistical Society*, Vol. 102, 1939, pp. 21-62.
7. L.D. Stamp, *Our Developing World*, Faber & Faber, London, 1960, pp. 104-110.
8. M. Shafi, Measurement of Agricultural Efficiency in Uttar Pradesh, *Economic Geography*, Vol. 36, 1960, pp. 296-305.
9. B.N. Ganguli, *Trends of Agriculture and Population in the Ganges Valley*, London, 1938, pp. 93-94.
10. S.G. Sapre and V.D. Deshpande, Inter-District Variations in agriculture, *Indian Journal of Agricultural Economics*, Vol. 19, No. 1, 1964, pp. 242-253.
11. S.S. Bhatia, A New measurement of agricultural efficiency in Uttar Pradesh *Economic Geography*, Vol, 43, No. 3, 1967, pp. 244-260.
12. M. Ali, Agriculture efficiency regions of Kashmir Valley, Paper Presented in the First Indian Geographical Congress, Section Regional Geography, New Delhi, 1972.
13. P.S. Gupta, Agriculture efficiency in food production in India Abstract of Papers, 21st International Geographical Congress, 1968 p. 166.
14. R. Chaudhury, et. al., *Soils of India*, ICAR, New Delhi, 1963, pp. 65-89.
15. R. Chaudhury, et. al., *Soils of India*, ICAR, New Delhi, 1963.
16. P.V. Sharma and S. Subrahmanyam, Rural development : Evaluation of some programmes, Paper Presented at the Seminar on Development of Rural Community, I.I.T. Kharagpur, April, 1976, pp. 3-8.

17. Ministry of Food and Agriculture, Government of India, Modernizing Indian Agriculture.
18. Fourth Five Year Plan, 1969-74, Draft Government of India, Planning Commission, pp. 106-107.
19. S.S. Acharya, Farming efficiency in relation to farm size and new techonology, *Yojana*, Vol. XVIII, No. 23, 1975, pp. 5-6.

12

Concepts of Innovation and Diffusion in Agricultural Land Use

THE PROCESS of agricultural diffusion is the spread of new varieties of seeds and implements in space over time. Agriculture is an operation of the cultivation of farm and the production of crops from land. The process of cultivation and production requires the help of human labour, mechanical implements, fertilizers, irrigation and pesticides. Innovation is a change in the established methods of agricultural practices, culture, ideas, etc. with the help of private intuition and the government machineries.

Any spread in diffusion process may be responded from only one centre or from differential location. However, the diffusion of a new variety of seed is dependent on the force generated by the place of diffusion, whereas the length of spread is the outcome of the mean information field of an adopter. Diffusion of any item depends upon the usability, popularity, ease of handling, source and force of information, level of income and the cultural traits of the adopter inhabiting the area. In this way, the diffusion, innovation and adoption of newer agricultural practices are related with the people involved in transportation, marketing and business, communications, advertisement, agricultural extension programme, rural sociology and several other scientific disciplines. In individual sector scientists approached the problem in terms of personal knowledge, social norms, economic condition and behavioural characteristics of people in adopting any new agricultural practices as an innovation. Geographers are interested in diffusion analysis due to its spatial and pattern analyses dimensions.

The need and importance of diffusion and innovation

studies for agricultural crops needs no emphasis. The recent start in population growth, without much talked improvement in agricultural production, it is unimaginable how one should be able to slove the food problem of millions of people living the South-East Asia. Further, it is hoped, this study will answer the following questions:

(a) What are the processes of innovations in which agricultural implements spread from one individual to another at varying locations?
(b) What are the inherent characteristics of innovators and the adopters?
(c) Why innovations are readily accepted by the farmers while certain others rejected?
(d) What is the role of agents and opinion leaders in the diffusion of innovations?
(e) What are the spatial patterns of diffusion of innovations?
(f) Which method should be adopted to analyse diffusion of innovations?
(g) Whether it is possible to predict the spatial spread of innovation?

The investigation about the diffusion of agricultural innovation has paramount importance in agricultural sector where the old society is in the process of change . The recent change in agricultural production has led to the economic growth and the growth of agro-based industries. Changes in the traditional agriculture system into a scientific one involves the following processes:

(a) agricultural research and innovation may remove the barriers of higher production, and (b) the level of the adoption of agricultural innovation by farmers according to the technological know-how and socio-economic conditions. In the field of agricultural research there are scientists who are capable of developing innovations but on the other hand, it is not certain that our farmers will readily accept it. In order to make available the fruits of scientific research and innovations to our ill-fed, ill-clad and ill-sheltered farmers, it is the duty of the scientist to make our farmers ready to accept the innovations of newer implements and agricultural practices.

REVIEW OF PAST LITERATURE

The diffusion study in geography dates back to early 20th century.[1] Among the pioneers, the work of Cooke presented a study of the diffusion of geographic phenomena,[2] i.e., plants, animals, birds and others. The original work of Hagerstrand's on Monte Carlo Simulation Model of spatial diffusion acted as catalyst for a number of studies carried on recently. Hagerstrand also pointed out that "a start is made by a rather concentrated cluster of carriers. This cluster expands step by step in such a way that the probability of a conversion always seems to be higher among those who live near the carriers than among those who live further away".[3] Further he[4] made a four stage model for the passage of innovation waves for dispersal of settlements. Firstly origin and evolution of initial agglomeration has been taken into account. Secondly diffusion is marked by centrifugal effect and creation of new centres in distant areas. Thirdly, in condensing process actual relative increase in the size of all locations is contemplated, while at the old stage asymptotic increase towards maximum dispersal is contemplated.

Nystuen (1963)[5] studied about the simulation of travel pattern in urban area. The effect of accessible and unaccessible barriers in space arising out from diffusion processes are analysed by Yuill (1964)[6] on urbanization. In the works of Morrill (1963)[7] analyses are available on simulation of urban fringe, and the growth of minority group ghettos in the city. In model alterations of land use at highway interchanges (Eighmy, 1964) along with major arterial routes leading out of a small centre have been employed (Colenutt, 1966) with the help of simulation methods. Most of the reviews on Monte Carlo Methods and geographical approach in spatial diffusion studies have been appeared about 1970's.

Bylund (1960)[8] has presented a four stage diffusion model of settlements in Inner North Sweden. It is an important contribution for dispersal process. The main differences among his four-stage model are the number and size of settlements increase and the location of original settlements. In this A and B models assumptions of spread are considered from coastal location while in C and D models from inland location of settlements. But these models are also applicable in Agricultural studies. In structural simulative model of settlements diffusion all occupance of the territory were shown through immigrational accretion, subdivision of clan groups into 'tappas' and 'parganas' boundaries and finally with the migration and competition, even distant areas have been filled up by the increase of varying sizes of settlements.[9]

In the same way Hudson (1969)[10] has investigated a new area of dispersal process in U.S.A. while he was investigating the settling process of Iowa. He has considered three processes colonization, spread and competition for the growth and dispersal of human habitations.

Processes of innovation and diffusion : Only two processes of innovation and diffusion i.e., relocation type and expansion type are quite discernible (Fig. 12.1). In relocation type some members of the population change their location from time to time *t-1* in space, while under expansion type new members are added between time *t* to *t-1*. Changes in the spatial patterns of diffusion depend upon six elements i.e., environment, temporal

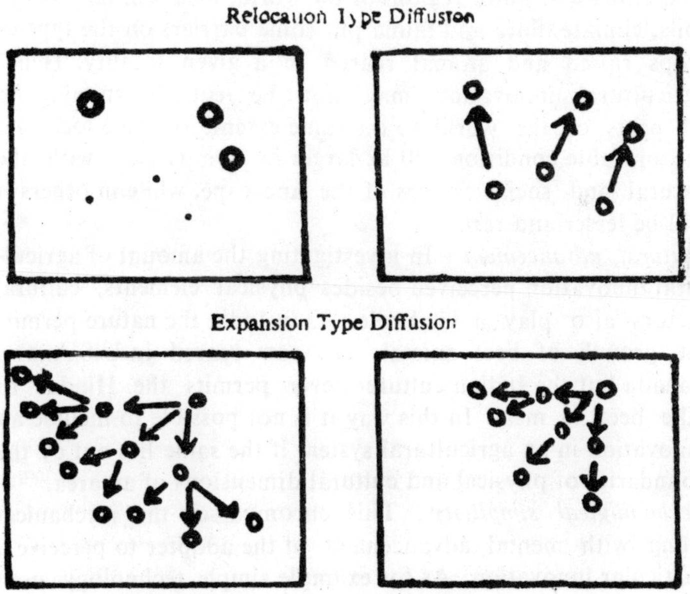

FIG. 12.1 : *Types of diffusion*

dimension, item being diffused, nodes of origin in space, nodes of destination at the end of time *t* and ultimately the path of movement in between origin and destination centres.

Dynamics of diffusion forces : Diffusion of crops focuses the clustering process of crops through innovation. This is dependent on the efficiency of communication system, technological and economic advancement of the society and the involvement of time in the propagation of innovation waves for diffusion of an item in space.

Hence, the factors of diffusion are innovation of crop, implements, channel of communication, social system, spatial process and the factors of time.

Innovation : Amongst several innovations only a few get diffused in an area. The causes are economic, cultural and perception of the adopters. The agricultural innovation of an area can be adopted by the farmers under the following conditions.[11]

Environmental feasibility : In any environment and technological availability a geographer can tell the suitability of crops grown. The exception of rugged relief features, forests and cold climate of Polar regions of the world, location, landforms, soils, climate, flora and fauna put some barriers on the type of crops raised and animal reared in a given locality. Hence agricultural innovation may not be equally suitable in all parts of the world to the same extent. In some localities the amicable condition will be larger in comparison with the natural and social factors of the landscape, while in others it will be lesser and rare.

Cultural advancement : In investigating the amount of agricultural innovation perceived besides physical elements, cultural factors also play a vital role. As in India the nature permits the growth of beef animals i.e. cow reared in U.S.A. and Canada but the Indian culture never permits the Hindus to take beef in meal. In this way it is not possible to include an innovation in an agricultural system if the same lies out of the boundaries of physical and cultural dimensions of an area.

Technological simplicity : This encompasses the mechanical along with mental advancement of the adopter to perceive a particular innovation. As for example simple technology used in plough can be easily acceptable to an Indian farmer than the complex tractor. This is due to technological and economic backwardness of the Indian farmers. The proper utilization of available natural resources also depends on technical abilities of peasants, because agricultural implements having complex device cannot be easily introduced to an uneducated fallow in comparison with the educated and financially well off people.

Experimentability : In a small plot of land, the experiment about the productivity of a crop could be easily ascertained after the division of large plot into smaller one. But it is difficult to plough smaller plot of land with the help of a tractor as it may not function if we make it part by part.

Communicability : Usually communicable innovations easily get diffused at wider locations than the stable and newly introduced items. As for example the seeds of several disease themselves get diffused in the air without the help of an adopter.

Economic feasibility : Uneconomic innovation has less chance to get diffused in space. As for example chemical fertilizer is more economical for the Indian farmers in comparison with combine harvester. This leads to the diffusion of chemical fertilizers quickly than combine harvester.

Reliability : Firm conviction of the diffusion of an idea is much helpful in the growth and development of agriculture. As for example in North Bihar BAC-32 a red and hard variety of sugarcane thrives well even in areas of lesser rain, flood conditions and against the menace of flood coupled with higher production of cane for crushing. In such cases of the adoption of innovation, risk is always involved there, as the investment may go infructuous in the purchase of seed, fertilizer and human labour if the physical environment is not suitable. For this fact many farmers in India are not desirous of adopting new variety of seeds as they are not in a position to take risk which involved in experiment through reliability on an item that the same will certainly get diffused.

Socio-technical desirability : Quite often farmers do not adopt all innovations which are economically gainful, but at the same time that should be desirable. Quite a large number of innovations do not spread in space as they are out of the socio-technical and political perimeters of the peasants. Large agricultural implements are not readily adopted by the peasants of India. The causes are not only ignorance of the machine but poverty and lack of repairing stations as such machines are almost absent from the traditional methods of farming in India. Hence there are three powerful forces—economic advantage, object poverty and past experience of traditional society which reject the diffusion of an innovation in space.

COMMUNICATION

In judging the diffusion of agricultural innovation the measure of mean information field of a farmer is highly helpful. According to Shanon and Weaver there are five basic elements in the communication process.[12]

(a) Source of information, (b) transmitter, (c) channel, (d) receiver, (e) destination.

The information may be available from a person or an organization. If the source about the innovation is reliable which may be of a crop or implement, the same may get diffused in shorter period with more reliability but in case of the suspicious authenticity of a news the transmitter as well as receiver will watch the situation before accepting or rejecting it. Normally an information passes through transmitter, channel, receiver and ultimately at a destination of an individual farmer or farm.

Social system : The intensity of innovation diffusion depend on the efficiency of transmitter and receiver under varying social and economic norms of a society. This is guided by the social 'tabus', values, age of a person and achievements in the field of agricultural operations. In a culturally advanced society an innovation will have to come across with a few barriers while in underdeveloped areas illiterate persons put hindrances in the diffusion of agricultural innovations.

Spatial system : Different parts of the earth surface have differential rate of the growth of agricultural crops or implements under innovation and diffusion processes. On sea coast, fishing industry will thrive easily while an unbridged embankment may hamper the process of diffusion equally because the physical conditions on two sides are not equal for the diffusion of an innovation. On different parts of the earth surface two types of effect of a barrier in transmitting the message have been realized e.g. due to the socio-spatial distance some message may increase or some may decrease and second time involved in transmitting the message. There are several causes of the maldistribution of diffusion of an information with the increasing distance and time in between the source of origin and destination.

RELATIONSHIP OF ADOPTION WITH DIFFUSION

In case any innovation fails to get adopted then neither adoption nor diffusion will take place. Hence it can be said that diffusion is impossible without adoption. According to Wilkening adoption is a process of learning, deciding, and acting over a period of time. The adoption of a specific practice is not the result of a single decision to act but of a series of actions and thorough decision.[13] Betal and Copp investigated five stages in adoption process:

(a) Awareness, (b) Attention, (c) Evaluation, (d) Trial, and (e) Adoption at the end.

One can divide the adopters into several types such as pioneers in investigating an idea, local, opinion leaders as early adopters, inquisitive followers who wait for the early adopters to demonstrate the usefulness of innovations before they adopt, sceptical followers who adopt an innovation under certain circumstances and laggards who are least concerned about the happenings which are going on in their physical and cultural set up.

In spatial aspects of adoption five stages can be included such as beginning of an idea, clustering, expansion, condensation and saturation at the end.

Mean information field : Mean information field denotes the spatial dimensions of knowledge about an information of a person on particular space and time. This is determined by the approachability of a person, transport connections, provision of communication lines, technical knowhow of adopters of a news and the rate of diffusion of an innovation, etc.

The mean information field data were collected from twenty-five farmers of Samastipur anchal of Samastipur district by random selection. The farmers were asked to give a list of persons with whom the contacts were generally made including the frequency of such contacts. In this way, it has been observed that most farmers know each other quite well within a radius of 3.5 kilometres of their own agricultural fields, but beyond that distance the pattern of contact varies from one individual to another. Usually these contacts decline from the centre of the farm to all directions uniformly while somewhere this rule may not follow. The reason for such biasness is due to the residence of the farmer on outskirts of the village, illiteracy, poverty, caste barrier, religious superstitions, etc., which would focus the interest of the farmers in a particular direction. The data collected from interview show a social contact with distance from the centre of the farm which is represented by the following equation.

$$P(D)=aD-b$$

where $P=$stands for number of contacts,
$D=$for distance from farmers home, and
$a=$and b are constants.

In any agrarian system of society the behaviour of a farmer is always probabilistic rather than deterministic. Thus in order to know the diffusion of agricultural innovations in space as a stochastic process we need a model to represent the normal observed tendencies. In diffusion analysis as distance between two cells increases the number of people who are known to a person decreases and this decrease is much faster than increase in a liner fashion. In this study the probabilities assigned to the different cells in probability field are derived from the pareto gravity potential function.

The idea underlies in this model is that human being have a close contact only in the nearby areas of a social system. In this sense we know that their relations in the form of close contacts are limited near the home and that can be easily analysed in a probabilistic model. Now let us assume that in an area of 25 square kilometres there emerges a transmitter of a given agricultural innovation. In this process of face to face transmission of information in the adjoining cells the number of people diffusing the information progressively increases because each additional adopter turns into a new transmitter also. The prior conditions for boosting diffusion are as follows:

1. The study should have an equally distributed population; 2. The transmitters do not go beyond four kilometres for pairwise meetings and tellings; 3. The potential adopters adopt the innovation as soon as they know of it; 4. All those who

adopt the innovation become transmitters; 5. Within a given generation, all persons adopt the innovation at the same time; 6. The transmitters move to any direction they wish for pairwise talks, and 7. There are no physical or cultural barriers against the movement.

Under these conditions the simulation of the movement of people for pairwise talk is possible, in order to develop sets of artificial data. This may be compared with the data observed from real world situation. This will indicate the success of the model and parameters used:

The above used equation could be written as:

$$\text{Log } P(D) = \log a - b \log D$$

Further, this relationship gets transformed into the following linear form after taking the logarithm of the number of contacts and distance on both sides:

$$Y = a - b\ x$$

where $Y = \log P\ D$

$X = \log D$

$a = \log a$

TABLE 12.1 : **Mean information field data from 25 farmers in the environs of Samastipur**

Distance from farmers home A in kilometre	Mean number of contacts per square kilometre for sample farmers
X	*Y*
0.50	10.27
1.00	9.12
1.50	8.32
2.00	7.25
2.50	5.50
3.00	3.20
3.50	2.12
4.00	1.97
5.00	0.55
27.50	49.13

$\overline{X}=2.75, \overline{Y}=4.91$

Thus in solving the regression equation

$Y=a-b\,x,$

b comes to 0.30541

while a comes to 0.22827

Therefore Log $(P)D$

$=0.22827-0.30541 \log D,$

where $P\,(D)=0.22827\,D-3.30541$

where $\log\,(PD)$ or $y=\log a+(-b) \log D$

with the help of $P\,(D)$ value, the number of contacts for each call of a five by five square grid of one square kilometre each was estimated according to its distance from the centre of the cell. For the central cell, actual number of contact (2) was used (Fig. 12.2).

Probabilities were obtained by summing the estimated contact values of each cell and dividing each by the total (5.552).

Thus, $Pi=\dfrac{Yi}{25}$

$i=1^{yi}$

192	.275	.317	.275	.192
.275	.544	.562	.544	.275
.317	.562	A	.562	.317
.275	.544	.562	.544	.275
.192	.275	.317	.275	.192

$\leq$ 11166

FIG. 12.2 : *Estimated number of contacts*

Where, ***Pi*** is the probability of contact between cell *i* and the central cell (*A*), and *yi* is the estimated number of contacts for cell *i* (Fig. 12.3a).

It is assumed that in all the above cells the number of potential adopters are the same. After considering Hagerstrand model, it could be said that the probabilities of transmitters getting the contact with receivers located in different cells will decline on an average at a rate which can be measured as reciprocal of the square of the distance. The probability field given for this purpose is given in Fig. 12.3b. Due to probability distribution it is necessary to add up the assigned values of different cells.

Later on in order to have a random selection of numbers, a range of discrete number is put in place of these probabilities (Fig. 12.3c). Now it is assumed that transmitter *A* is located in one of the cells of the probability field. In g^{o} *A* makes a move for pairwise meeting in all directions around him. Now in the selection of random number of three digits

016	023	.027	.023	016
023	046	.048	046	.023
027	.048	A	.048	.027
.023	.046	.048	.046	.023
.016	.023	.027	.023	.016

a

FIG. 12.3a : *An estimate of simulation*

An Estimate of Simulation

16	17-39	40-66	67-89	105
106-128	129-174	175-222	223-268	269-291
292-318	319-366	A	367-414	415-441
442-464	465-510	511-558	559-604	605-627
643	644-666	667-693	694-716	732

FIG. 12.3b

An Estimate of Simulation

1				
2	4	7		
3	5	9 A	6 C	4 F
		5 B	4 D	3 G
		3	?	1

FIG. 12.3c

from left to right move, suppose we get 417 and the same is out of our cells, so it should be ignored. Suppose our next number is 449 which is within the field and is located one cell below the *A's* cell. Now one more person knowing the massage may call *B*; and here is end of generation one (g^1). In this generation both *A* and *B* are transmitters, and now both of them convert one more person as transmitter, so up to g^2 there will be four transmitters i.e., *A,B,C* and *D*. This process may continue in order to see the further situations. In this analysis the transmitters were permitted to move in any direction within the field.

TRENDS IN DIFFUSION STUDIES

Recently the innovation and diffusion processes from the subject matter of several scientific discipline. In this regard sociologists analyse the social structure, group behaviour and farm technology with particular reference to the impacts of community norms, social customs and cultural values, etc. The psychologists stressed on problem of human perception; regional scientists studies about the spread of diseases and the zone of influence of hospitals; while economists analysed the spread of innovation from one profitable location of a firm to another. Geographers are mainly concerned with the adoption and spatial dimensions of technological innovations, impact of new transportation facilities near settlements, limits of population migration, seeds, fertilizers, agricultural implements, cultural values, spread of diseases and epidemics, limits of marriage connections and human perception in diffusion of innovation studies. Further geographers are interested in the differential processes of the spread of different items over space and in finding out the nature of barriers which obstruct the communication of ideas of innovations in order to predict the rate, so that other innovations or ideas may spread over space under similar conditions. Hence geographers have to study diffusion problem in comparison with sisters disciplines and their contributions are mostly directed towards inter-disciplinary approach in solving many diffusion problems.

The geographers approach to diffusion may be summarized as:

(a) the diffusion of any item tends to wane with the increasing distance as it travels from one place to the other and one individual to another.
(b) spatial diffusion is determined by the force generated by a message after propagation from its origin point.
(c) distribution of any item or its location is partly determined by a chance factor.
(d) some areas lag behind others in the adoption of innovations, and
(e) the hindrance in the migration of agricultural crops are functionally different in nature according to conditions of terrain, climate, social and economic wealth of a region.

REFERENCES

1. E. M. Rogers, *Diffusion of Innovations*, Free Press, New York, 1962.
2. M.T. Cooke, The Spread of the European Starling in North America, U.S. Department of Agriculture, 1928.
3. T. Hagerstrand, On Monte Carlo simulation of diffusion (Elt.W. Garrison), *Quantitative Geography*, Department of Geography, Northwestern University, Evanston, 1963, pp. 1-32.
4. T. Hagerstrand, The propagation of innovation waves, *Lund Studies in Geography*, Series-B, *Haman Geography*, No. 4, 1952.
5. J.D. Nystuen, an approach to the direct measurement of community mean information fields, *Papers and Proceedings of Regional Science Association*, Vol. II, 1963, p. 99.
6. R.S. Yuill, A Simulation study of barrier effects in spatial diffusion problems, *Spatial Diffusion Study*, Technical Report, No. 1, 1969.
7. R.L. Morrill, The distribution of migration distances, *Papers and Proceedings the Regional Science Association*, Vol. pp. 75-88.
8. E. Bylund, Theoretical considerations regarding the distribution of settlement in inner North Sweden, *Geografisca Annaler*, Vol. 42. pp. 225-231.
9. K.N. Singh, The territorial basis of medieval town and village settlements in Eastern Uttar Pradesh, *Annals of the American Geographers*, Vol. 58, No. 2, 1968, pp. 203-229.
10. O.C. Hudson, Diffusion in a central place system, *Geographical Analysis*, Vol. I, 1969, p. 46.
11. R.P. Misra, *Diffusion of Agricultural Innovation*, University of Mysore, 1968.

12. C.E. Shanon and W. Weaver, *The Mathematical Theory of Communication*, Urbana, University of Illinois Press, 1949.
13. E.A. Wilkening, Social isolation and response of farmers to agricultural programme, *American Sociological Review*, 16, 1951, pp. 836-837.

13

Methods of Crop Combination

THE CONCEPT of crop combination is a scientific devise to study the existing spatial relationship of crops in association with each other in agricultural geography and land utilization. Such study is necessary in order to have a more complex structure of agricultural regions. The measures the areal dominance of crops which are closely related and which occur together in varying strength.[1] On lower-level of crop dominance diversities and complexities persist, but this can easily be relegated and studied with limitation of varying crop associates[2] according to their importance in the region. This idea of crop combination could be best utilized in understanding the existing agricultural situations, the land used patterns of a region and the planning of crops for the future. Until now several methods of crop combination have been presented e.g. decreasing order arrangement of the percentage of different crops followed by the calculation of Standard Deviation for ranking of different set of crops by Weaver (1954)[3]; maximum distance method of Ayyar (1969)[4]; method of difference and summation and lower limit method of Athawale (1966)[5]; Standard Deviation method by Betal (1976)[6] after Weaver but not even a single method is presented for preserving the integrity of higher percentange of crops grown in a region and simple calculation. Hence after making a review of number of methods this chapter has been designed to devise a new method to study the crop combination in the State of Bihar as a case study and it can be applied universally. The data used here have been collected from the Department of Statistics and Evaluation Government of Bihar for the year 1970-71.

METHODS OF CROP COMBINATION

In order to have a more complex structure of agricultural regions the study of crop combination is necessary.

Method 1

Weaver[7] the pioneer of quantitative idea on crop combination has considered a decreasing order arrangement of the percentage of different crops such as :

A=50 and B=50

In the same way A=33, B=33, C=33, and A=52, B=28, C=12, D=8.

In these arrangements the combinations are AB, ABC as unambiguous ones but last combination is ABCD and, as the decreasing order percentage, it is an ambiguous combination.

In this way Weaver's standard theoretical distributions are as follows:

100.0% of GCA for monoculture i.e., single crop
50.0% of GCA for each crop i.e., two crop combination.
33.3% of GCA for each crop i.e., three crop combination.
25.0% of GCA for each crop i.e., four crop combination.

Where GCA is gross cropped area. The deviations (d) plus or minus of the actual percentage from any one theoretical values for every set of crops are squared and summed. Each sum is divided by the number of crops inthe set. The square root of this result will give us a measure of crop combination from the theoretical distribution. The formula for Standard Deviation is as follows:

$$\sigma^2 = \frac{\Sigma (d^2)}{n}$$

Where σ^2= Standard Deviation
d^2= deviation from mean square
n = number of crops
Σ = summation value of d^2 in an areal unit.

Weaver considered all the crops having at least one per cent acreage of the total cropped land for the analysis.

The main objections against this method is as follows:

(a) The time involved for all these calculations is too much.
(b) The crop combination cannot be determined till all the *n*th crops are considered.

(c) Sometimes quite a large number of crops are included in the crop combination.

Example : Percentage distribution of the average area for 1970-71 under 7 different crops in Monghyr district of Bihar is given below, using Weaver's method to identify the significant crops of the district.

% of Crop	Maize	Wheat	Rice
Area	28	22	21
Gram	Arhar	Barley	Sugarcane
7.28	2.35	2.24	0.70

Solution

ONE CROP

			Summation
Hypothetical %	100		
Observed %	28		
Difference, d	72		
Σd^2	5184		=5184
$\frac{\Sigma d^2}{n}$	5184		5184

TWO CROPS

			Summation
Hypothetical %	50	50	
Observed %	28	22	
Difference, d	22	28	
Σd^2	484	+784	=1268
$\frac{\Sigma d^2}{n}$			634

THREE CROPS

				Summation
Hypothetical %	33.3	33.3	33.3	
Observed %	28.0	22.0	21.0	
Difference, d	5.3	11.3	12.3	
Σd^2	28.09	+127.69	+151.29	=307.07
$\frac{\Sigma d^2}{n}$				102.36

FOUR CROPS

Hypothetical %	25	25	25	25.00	Summation
Observed %	28	22	21	7.28	
Difference, d	3	3	4	17.72	
Σd^2	9	+9	+16	+314.00	=348
$\frac{\Sigma d^2}{n}$					87

FIVE CROPS

Hypothetical %	20	20	20	20	20	Summation
Observed %	28	22	21	7.28	2.35	
Difference, d	8	2	1	12.72	17.65	
Σd^2	64	+4	+1	+161.80	+311.52	=542.3
$\frac{\Sigma d^2}{n}$						108.46

The district may, therefore, be identified as a four crops region (as the value of the index is least for four crops) having Maize as the most dominating crop followed by wheat, rice, gram and arhar respectively.

Thomas (1963) modified Weaver's method by including the crops with zero per cent theoretical values also in each step of comparison. As this method involves all the crops in each step there is no need of dividing Σd^2 by n.

Example : As an example the crop combination of Singhbhum district has been identified using the modification suggested by Thomas.

ONE CROP

Hypothetical %	100	0	0	0	0	0	0	Sum-
Observed %	84	3	2	1.25	.40	.25	.16	mation
Difference, d	16	3	2	1.25	.40	.25	.16	
d^2	256	9	4	1.56	.16	.06	.03	=270.81

TWO CROPS

								Sum-mation
Hypothetical %	50	50	0	0	0	0	0	
Observed %	84	3	2	1.25	.40	.25	.16	
Difference, d	− 34	47	2	1.25	.40	.25	.16	
d^2	1156	2209	4	1.56	.16	.06	.03	=3370.81

Since the value of crop combination is increasing towards higher number of crops hence it is a case of mono-crop region, and rice is the dominant crop.

Method 2

Doi's Method : Doi has defined in detail his statistical procedure in his paper the 'Industrial Structure of Japanese Perfecture'. His modified formula is that of Weaver is as follows :

$$\Sigma d^2$$

According to this formula the crop combination is as follows :

1. Monoculture $=(100\text{-}35)^2=4225.00$
2. Crop combination $= (50-35)^2+(50-24)^2 = 901.00$
3. Crop combination $= (33.33-35)^2+(33.33-24)^2-33.33-15)^2=423.70.$

Similarly the other combinations may be computed.

Since the variance for three crop combination (423.70) is smallest, the country is designated with a combination of three crops, i.e; WBP.

On deriving crop-combination for unit of X' crops on Doi's method, it was found that the results are more realistic in comparison with those worked out by other methods. It is easy in calculation within a short time.

Method 3

Rafiullah Method : Rafiullah introduced new deviation formula which although suffers from the defects of huge calculations is quite suitable for delineating sharply the primary crop combination. The formula is as follows:

$$d=\sqrt{\frac{DP^2-D^2n}{N^2}}$$

Where d is the deviation

DP is the positive difference

Dn is the negative difference from the mean value of the theoretical curve value of combination.

N is the number of crops in the combination.

Since only relative rar.k of the deviation is needed, the root sign may be ignored to save calculations formula may be used as :

$$d^2=\frac{\Sigma DP^2-\Sigma Dn^2}{N^2}$$

Unlike Weaver difference of actual values is calculated from the middle value of the theoretical standard and the maximum positive deviation gives the critical combination.

Method 4

Maximum distance method : Deviating from Weaver's idea of crop combination N.P. Ayyar has tried to simplify such investigation and his grouping method is as follows :

all cereals	(7)	.... *C*
all pulses	(3)	.... *P*
all oil seeds	(6)	.... *O*
all other food crops	......	.... *F*
all other non-food crops		.... *N*

The figures in brackets show the number of crops in each group while *C, P, O, F, N,* indicate the symbol for each group. Thus the standard values may be obtained which are as follows:

C 100.0%

C-P...............70.0%—30.0%

C-P-O...........43.7%—18.8%—37.5%.

C-P-O-F..........31.8%—14.3%—28.6%—23.8%

C-P-O-F-N......26.9%—11.6%—23.1%—19.2%—19.2%.

The deviations of the actual percentages from the above values would be found out, squared and summation (Σ) will give σ^2 (standard deviation) as required by the crop group combination. Now the method of difference, method of summation and lower-limit method have been used here in words of A.G. Athawale.[8]

Method 5

A method of difference : In case of greater difference between first and second ranking crop than the second crop is not important and this will be a case of monoculture. If this difference is little and the difference between second and third ranking crops is much, than the third crop is not important and it is a case of biculture and so on. The solution of obtaining high or low values are as follows:

(a) With higher difference between the percentage of 1st and 2nd ranking crops i.e. $\frac{100}{1}-\frac{100}{2}=50$. In this case first ranking crop is important and the combination would be monoculture.

(b) In relation to theoretical difference if the difference between second and third ranking crop is higher i.e. $\frac{100}{2}-\frac{100}{3}=16.67$ and this is the case of bi-culture.

(c) When the difference between the percentages of nth and $(n+1)$ the ranking crops is more than the difference between the theoretical values i.e. $\frac{100}{n}-\frac{100}{n+1}=\frac{100}{n(n+1)}$ then the crop combination would consist of first n ranking crops.

By inserting different values to n in $\frac{100}{n+(n+1)}$ a table can be prepared. Then by taking actual differences and referring those to table values the crop-combination can be determined immediately.

TABLE 13.1 : Crop combination

n	1	2	3	4	5	6	7	8	9
$\frac{100}{n(n+1)}$	50.00	16.67	8.33	5.00	3.33	2.33	1.79	1.39	1.11

Example 2: Let values of X_1, X_2 and X_3 be 70, 25 and 5 respectively. The difference between 1st and 2nd ranking crops is 45 which is less than the corresponding table value for $n=1$.

The second difference is 20 which is more than table value for $n=2$. Hence this is a case of bi-culture crop combination.

Example 3: Let value of X_1, X_2, X_3, X_4, and X_5 be 54, 24, 13, 5, and 2 respectively. The various differences are 30, 11, 8, 3 and 2 respectively none of which is more than the corresponding table value. When such indeterminate results are obtained it is suggested that the case may be treated as all crop combination.

Method 6

Method of summation : In this method instead of observing the differences, progressive sums are taken. The actual procedure is explained below:

(a) If the percentage of first ranking crop is more than the average of theoretical value for 1st and 2nd crops, i.e. more than $\frac{1}{2}\left(\frac{100}{1}+\frac{100}{2}\right)=75$ then it is a case of monoculture.

(b) If the sum of percentage of first two ranking crops is more than twice the average of theoretical values for 2nd and 3rd ranking crops i.e. more $2\times\frac{1}{2}\left(\frac{100}{2}+\frac{100}{3}\right)=83.33$, thus it is a case of bi-culture.

(c) If the sum of percentages of first n ranking crops is more than n times the average of theoretical values for nth and $(n+1)$ the crops i.e. more than $n\times\frac{1}{2}\left(\frac{100}{n}+\frac{100}{n+1}\right)=$

$$\left(\frac{2n+1}{n+2}\right)100$$ thus it is a case of n crop combination.

By preparing table for $\left(\frac{2n+1}{2n+2}\right)100$ by giving different values to n it is possible to determine the crop combination just by taking progressive sums and referring to table values.

TABLE 13.2 : Crop combination taking progressive sum

n	1	2	3	4	5	6	7	8	9
$\frac{(2n+1)}{(2n+2)}$	75	83.33	87.50	90	91.67	92.86	93.75	94.44	95

Example 4: Let values of X_1, X_2 and X_3 be 70, 25 and 5 respectively. First cumulative sum is 70 which is less than

table value for $n=1$ and hence crop combination is not monoculture. The second cumulative sum $(70+25)$ is more than the table value for $n=2$ so it is a case of bi-culture crop combination.

Method 7

Lower limit method : The above method of taking differences or progressive sums is discontinued after getting the result. If the value of X_1, X_2, X_3 and X_4 be 76, 12, 10 and 2 respectively then by method 1 and 2 the result is monoculture crop combination even though the percentage acreage of second and 3rd ranking crops are quite high. Of course, by any method such marginal cases are bound to occur. In lower limit method an attempt is made to associate the value of N to determine the crop combination. This is achieved in the following way which gives the lower limit for percentage values.

$$P=\frac{100}{3N}$$

The crops having percentage acreage more than P will be included in the crop combination. By giving different values to N a table of P values can be:

TABLE 13.3 : **Crop combination having percentage acreage**

N	1	2	3	4	5	6	7	8	9
P	33.33	16.67	11.11	8.33	6.67	5.56	4.76	4.17	3.70

Example 5: Let X_1, X_2, X_3 and X_4 be 76, 12, 10 and 2 respectively. The value of P for $N=4$ is 8.33.

So all the crops having percentage acreage more than 8.33 will be included in the crop combination. There are 3 crops above this limit. So it is a three crop combination.

As we have seen above the result is monoculture crop combination by method 1 and 2. It is interesting to note that by Weaver's method also it comes to monoculture. So the result detained by method 3 is more appealing. Also the saving of time is considerable. In addition to the advantage gained in time factor one of the greatest advantage of this method is its flexibility. If broad regions are required then instead of $100/3n$

TABLE 13.4 : Monoculture crop combination

District	1	2	3	4	5	6	7	8	9	10	11	12	Method 1	Method 2	Result Mod 3 100/3n	Result Mod 3 100/2n
Anantapur	22.5	18.5	15.9	10.9	7.4	7.2	6.9	4.2	2.6	1.6	1.1	1.1	7	9	8	8
Chitoor	31.1	24.9	12.0	10.1	6.7	5.4	4.2	3.7	1.4	—	—	—	8	7	8	5
Cuddapah	20.2	19.9	16.2	14.0	7.1	6.4	6.1	4.1	3.9	1.2	—	—	4	9	9	7
East Godavari	60.5	8.4	8.0	4.5	4.2	4.1	3.8	1.6	1.6	1.3	—	—	1	8	7	3
Guntur	22.5	19.6	14.7	9.6	8.7	7.9	5.3	5.2	4.3	1.0	—	—	6	9	9	8

one can take 100/2*n* or 100/*N* or any multiple of *N* in the denominator to decide the lower limit.

In all the methods discussed above the percentages are first to be calculated. The part of calculations also take a lot of time and it can be saved by modification in Method 3 which is as follows:

$$A=\frac{G}{3N}$$

Where G=Gross cropped area

N=number of crops having more than or equal to $G/100$ i.e. 1 per cent.

A is the lower limit in terms of acreages. Then all crops having acreage more than A will find place in crop-combination. An example is as follows:

When $G=1260$ and the crop acreages be 800, 200, 160, 50, 25, 10, 10, 10 and 5 than $G/100=12.60$.

Hence $n=5$

and lower limit $A=G/3N=70$.

There are three crops above 70 and hence combination would be triculture.

This idea of lower-limit method is best due to its simplicity and flexibility, but the diversification of crops may arise mapping problem.[9]

Method 8

Betal Himansuranjan, while studying the problems of agricultural growth in India sets out Weaver's method of crop combination after taking mean and standard deviation of different crops for the states of India. The calculation for this method is giver here for rice crop in Bihar State.

TABLE 13.5 : **Mean and standard deviation of different crops**

Districts	%acreage under rice crop	Deviation from mean	Deviation squared
1. Patna	39	−11	121
2. Gaya	53	+3	9
3. Shahabad	49	−1	1
4. Saran	30	−20	400
5. Champaran	50	000	000
6. Muzaffarpur	42	—8	16

(*Contd.*)

TABLE 13.5 (*Contd.*)

7. Darbhanga	48	—2	4
8. Monghyr	21	—29	841
9. Bhagalpur	40	—10	100
10. Saharsa	44	—6	36
11. Purnea	62	+12	144
12. Santal Parganas	69	+19	361
13. Hazaribagh	56	+ 6	36
14. Ranchi	62	+12	144
15. Palamau	30	—20	400
16. Dhanbad	78	+28	784
17. Singhbhum	84	+34	1156
Total (17)	857		4553

(Mean) $\bar{X}=50\%$

Then the calculation of Standard Deviation is as follows:

$$\sigma^2=\sqrt{\frac{\Sigma(d)^2}{n}}$$

$$=\sqrt{\frac{4553}{17}}$$

$$=\sqrt{268}$$

$$\sigma^2=16.37$$

Then the range is as follows:

Rank III—Mean+1=50+16.37=66.37

Rank II —Mean+2=50+66.37=82.74

Rank I —Mean+3=50+82.74=99.14

Then for different districts the following ranks for the rice crop are determined.

Singhbhum	—I
Dhanbad	—II
Santal Pargana	—II
Gaya	—III
Champaran	—III
Purnea	—III
Hazaribagh	—III
Ranchi	—III

In this way the worksheet for the major crops are computed to facilitate the crop combination. The results are tabulated below:

TABLE 13.6 : Worksheet for the major crops are computed to facilitate the crop combination

Major Crops	Symbols	Mean ($\overline{X}$)	Standard Deviation (σ)	Ranks of Distinctiveness ($\overline{X}$)+3	($\overline{X}$)+2	($\bar{X}$)+2
				1	2	3
Rice	R	50	16.37	99.14—82.74	82.74—66.37	66.37—50.00
Sugarcane	S	1.33	1.53	5.97— 4.44	4.44—2.91	2.91—1.38
Marua	Ma	1.86	2.06	8.04—5.98	5.98—3.92	3.92—1.86
Wheat	W	10.38	7.66	33.36—25.70	25.70—18.04	18.04—10.38
Arhar	A	1.52	1.10	4.82—3.72	3.72—2.62	2.62—1.52
Maize	M	10.23	6.87	30.84—23.97	23.97—17.10	17.10—10.23
Gram	G	2.32	1.70	7.42—5.72	5.72—4.02	4.02—2.32
Barley	B	2.24	1.90	7.94—6.04	6.04—4.14	4.14—2.24

Source : Calculated by the author from Season and Crop Report, 1970-71.

On the basis of the scale of distinctiveness the crop combinations can be computed for the different districts of Bihar State.

TABLE 13.7 : **Distinctiveness the crop combination can be computed for the different districts**

Districts	Crop Combination		
	1st Rank	2nd Rank	3rd Rank
Patna	—	—	R, S, W, G
Gaya	—	—	R, S, W, G
Shahabad	—	W	G
Saran	S, A, B.	M	W
Champaran	S	B	R, W
Muzaffarpur	—	B	S, W, A, M
Darbhanga	—	Ma	S,W, A, M
Monghyr	M,G.	W	A, B
Bhagalpur	—	—	M, G, B
Saharsa	—	Ma	W, M
Purnea	—	—	R, W
Santal Pargana	—	R	S
Hazaribagh	Ma	A	R, M
Ranchi	Me	—	R, A, G
Palamau	—	B	M
Dhanbad	—	R,G.	Ma, A
Singhbhum	R	—	—

Source : Compiled by the author.

With the above two tables of crop combination it has been revealed that higher percentage of crops have come under the third rank and lower percentage of other crops on the first rank. This is the main defect of this method of crop combination.

New method :

A new method has been envisaged here for crop combination on regional level. The formula is as follows:

$$C.C. = \begin{cases} \dfrac{HP}{1} - 1 = \text{Ist ranking crop} \\ \dfrac{HP}{2} - 2 = \text{IInd ranking crop} \\ \dfrac{HP}{3} - 3 = \text{IIIrd ranking crop} \\ \dfrac{HP}{4} - 4 = \text{IVth ranking crop} \end{cases}$$

Lower limit for crop diversification.
Where CC is crop combination.

HP is the highest percentage of a crop in the study area.

1, 2, 3, 4 are denominators for crops of a particular rank.

and —1,—2,—3,—4 are diversity index for crops of low percentage.

Computation of crop combination on the basis of above formula is as follows:

Singhbhum District

Rice 84%	Sugarcane 0.25%
Arhar 1.25%	Marua 0.16%
Maize 3%	Wheat 0.40%
Gram 2%	

84/1-1=83% Ist ranking crop.
84/2-2=42-2=40% IInd ranking crop.
84/3-3=28-3=25% IIIrd ranking crop.
84/4-4=21-4=17% IVth ranking crop.
Rice (Ist ranking crop).
A case of monoculture.

An other example is also presented here for multiple crop combination.

Monghyr District

Rice 21%	Barley 2.24%
Wheat 22%	Arhar 2.35%
Maize 28%	Sugarcane—0.70%
Gram 7.28%	

28/1-1=27% Ist ranking crop.
28/2-2=14-2=12% IInd ranking crop.
28/3-3=9.33-3=6.33% IIIrd ranking crop.
28/4-4=7-4=3% IVth ranking crop.

I	II		II	III	: Multiple Crop
Maize	Wheat	—	Rice	Gram	Combination.

TABLE 13.8 : **Ranking of crops in Bihar (1970-71)**

District	Ranks of Distinctiveness according to the percentage of crops			
	I	II	III	IV
Patna	38 Rice	38-17.5	17.5-10 Wheat	10-5.75 Maize
Gaya	52 Rice	52-24.50	24.50-14.66 Wheat	14.55-9.25
Shahabad	48 Rice	48-22.50	22.50-13.33 Wheat	13.33-8.25
Saran	29 Rice	29-13 Maize, Wheat	13-7 Barley	7-3.50 Sugarcane Arhar
Champaran	49 Rice	49-23	23-13.66 Wheat	13.66-8.50
Muzaffarpur	41 Rice	41-19	19-11 Maize, Wheat	11-6.50
Darbhanga	47 Rice	47-22	22-13 Wheat, Maize	13-8
Saharsa	43 Rice	43-20	20-11.66 Wheat, Maize	11.66-7
Purnea	61 Rice	61-29	29-17.66	17.66-11.50
Monghyr	27 Maize	27-12 Wheat, Rice	12-6.33 Gram	6.33-3
Bhagalpur	39 Rice	39-18 Maize	18-10.33 Wheat	10.33-6
Palamau	29 Rice	29-13	13-7 Maize	7-3.50 Barley, Wheat
S.P.	68 Rice	68-32.50	32.50-20	20-13.50
Hazaribagh	55 Rice	55-26	26-15.66 Maize	15.66-10
Ranchi	61 Rice	61-29	29-17.66	17.66-11.50
Dhanbad	77 Rice	77-37	37-23	23-15.50
Singhbhum	83 Rice	83-40	40-25	25-17

Source : Compiled by the author.

The above method of crop combination has advantage over several other methods in two respects. Firstly, it takes less time in calculations and secondly, it maintains the integrity of higher percentage of crops grown in a region. For instance, in Betal's method of crop combination through Standard Deviation, lower percentage value emerges as 1st ranking crop over highest percentage of other crops in a region. But this new method relegates such lacuna so far as the methodological efficiency is concerned. The other lacuna in Betal's method is that lesser percentage of crops emerges for combination with the dominant crops, but such association may not persist for a longer period, because less important crop in a region never represents the agricultural situation and ecological condition of a region.

First ranking crops : The State of Bihar comes under rice belt of India and naturally, rice is a predominant crop in all the district except Monghyr where maize dominates. Table 13.5 gives an idea that Purnea, Singhbhum, Dhanbad, Ranchi and Santal Pargana districts have first ranking monoculture cultivation of rice. In these areas, this crop has percentage share of 62 per cent in Ranchi and Purnea and 84 per cent Singhbhum district. Owing to heavy monsoonal rain rice is an important crop, but in other seasons the intensity of cultivation is low intervening with the forest environment. In these areas the agricultural situation is mostly affected by the availability of rainfall in the Don Lands or Valleys while Tenure land or higher ground is mostly utilized for terraced cultivation of rice. In this uneven terrain of Chota Nagpur plateau, other crops have very poor share in the general scheme of land utilization, and for their fact the method of crop combination relegates such crops to association. On the other hand, in the districts of northern plains rice comes as first ranking crop due to higher percentage of rice in comparison with other crops.

Second ranking crops : In Saran district maize (19%) and wheat (18%) come under second ranking crops, while in Monghyr district wheat (22%), rice (21%) and in Bhagalpur district maize (17%) come under second ranking crops. This is due to the predominance of rice culture everywhere and no crop has very near alternative to the rice in this area of sedentary cultivation.

Third ranking crops : Except uneven terrain of Chota Nagpur plateau everywhere wheat is a predominant third ranking crop in the State of Bihar and it is followed by barley, maize and gram. The share of land under this category in different districts is as follows: In Saran barley 7 per cent, in Darbhanga district wheat and maize 12 per cent, in Monghyr district gram 7 per cent, in Bhagalpur district wheat 10 per cent, in Gaya district wheat 14 per cent, while in Muzaffarpur and Saharsa district wheat and maize are important third ranking crops—occupying 12 per cent, 14 per cent, and 13 per cent land respectively.

Fourth ranking crops : The crops which have one-fourth percentage of the first ranking crop mines four as diversification index come under this category. In Palamau district barley 4.63 per cent and wheat 4.35 per cent in Patna district maize 8 per cent, in Saran district sugarcane 4.50 per cent and arhar 4.50 per cent come under fourth ranking crops.

CONCLUSION

The concept of crop combination should have clear image of association of crops with decreasing intensity of crop land use and its regional economic significance, so that the ills of low percentage crops can be removed by technological advancement, fertilization, conservation of soil, by controlling weeds, pests and flood menace, and with the provision of irrigation and better seeds.

Recently, governmental measures have put forward several schemes for all round development of agricultural land. With these achievements, the State of Bihar is marching forward in attaining self-sufficiency in food production in the very near future and hence it is worthwhile to have a record of our crop association to know the future trend and agricultural planning. In this way this paper spells out a new method of crop combination in the fields of land utilization, agricultural geography and planning.

REFERENCES

1. E. Dayal, Crop combination regions : A case study of the Punjab Plain, *Tefdschrift Voor Economische on Sociale*, Vol. LVIII, 1967, No. 1, p. 39.
2. B. Mandal, Crop-combination regions of North Bihar, *The National Geographical Journal of India*, Vol. XV, Part-2, 1969. p. 125.
3. J.C. Weaver, Crop combination regions in the Middle West, *Geographical Review*, 1954, Vol. 44, pp.1 75-200.
4. N.P. Ayyar, Crop regions of Madhya Pradesh : A study in Methodology", *Geographical Review of India*, Vol. XXXI, No. 1, March, 1969, pp. 4-5.
5. A.C. Athawale, Some new methods of crop combination, *Geographical Review of India*, 1966, Vol. XXVIII, No. 3, pp. 28-34.
6. H. Betal, *Crop Combination Regions of India*, *Geographical Aspects of Indian Agriculture*, Department of Geography, Calcutta University, 1976, pp. HRB 1, to HRB 7.
7. J.C. Weaver, op. cit.
8. A.C. Athawale, Some new methods of crop combination, *Geographical Review of India*, 1966, Vol. XXVIII, No. 3, pp. 28-34.
9. H. Betal, Crop combination regions of India", *Geographical, Aspects of Indian Agriculture*, Department of Geographyical, Calcutta University, 1976, pp. HRB 1 to HRB 7.

14

Land Resource Measurement

INTRODUCTION

THE LAND resource can be measured by several methods. It may be assessed in five ways i.e., (a) output per unit area, (b) output per worker, (c) the value of agricultural output, (d) input-output ratio and (e) Per-capita output of grain. In India we have scanty land resource, hence the production per unit of area is more important than the production per worker. It is, therefore, possible to plan maximum use of our land but the same is not consistent with the labour and capital involved. In indian agriculture, two types of labour are used: human and draught cattle. In such a situation, human labour may be cheap as India is a populous country, but the cattle labour is inadequate.

However, the produce of the land or land resource should be measured in terms of 'man', because land is the chief source of sustenance for all lives. The land itself is controlled by its physical, chemical and biological properties, the environmental features of a particular latitude, altitude, climatic conditions and rainfall which are the prime determinants of a particular terrain condition. The various types of land so conditioned are best suited for a particular use of man. In several countries of the world, the scientific discoveries of man has ruined even the original characteristics of land given by the nature. With the advancement of engineering works man has now profound control over water, moisture, soil features, climatic conditions, science of forestry, checking the advancement of deserts by soil conserving grasses and by planting wind breaking forest fringes, etc. In this way man has been able to determine the nature of land utilization with various degrees of success. Besides this, the development of soil chemistry, soil bacteriology, plant chemistry and plant genetics has opened a new horizon for successful economizing in the use of land and supplementing our food with the exploitation of food resources e.g., fish and algae from seas, rivers, lakes and other forms of water bodies.

A MODEL STUDY

Prof. Stamp used "to determine the farming efficiency of twenty countries on the basis of per acre yield of nine selected food crops. He had taken care of certain crops only which were grown in nearly all the countries, but in case of certain crops only averages were used. The countries were placed in order of the output per acre of each crop. The places occupied by each country in respect of the selected crops were then averaged and from these averages the ranking co-efficient of agricultural efficiency of each country was obtained. For example, if a country was at the top of every list, it would have a ranking coefficient of one, and if it were at the bottom of every list, it would have a ranking co-efficient equal to the total number of countries concerned".[1]

Huntington and Valkenburg used a different method to measure the intensity of agricultural production. "They considered the productivity of land per acre on the basis of eight very widely raised crops in Europe. For each crop, the average yield per acre for Europe was taken as an index of 100, and the specific yield in each country was calculated accordingly. Prof. Stamp pointed out that the fundamental difficulty in measuring food production efficiency in different parts of the world or in any sizeable area is the range and variety of crops. It may not be fair to equate a pound of wheat with a pound of millet as their caloric yields are vastly different. Of course, the difficulty can be overcome by converting the production of any crop into nutrition calories. This eliminates the variations in quality and provides a common basic of assessment".[2]

In order to accept the last method of 'Per-capita output of grain', J.L. Buck[3] devised method as output expressed in terms of 'grain equivalent'. He has considered all food-grains to be equivalent in food values and measured in terms of per kilogram, which may either be produced locally or purchased from outside.

A modified method of Vries considered output in terms of 'milled rice equivalents' per-capita. He converted all food-grains equal to rice equivalents according to the local market price of various grains. This method will not be applicable especially in the multi-crop regions as there may not be single diet much more prevalent in the dietary habit of the people. A slight modification of the procedure has been introduced by Clark and Haswell who expressed the agricultural output in terms of kilograms of wheat equivalent per person. For this purpose they employed a scale based on the weighing system used by the Food and Agricultural Organization of the United Nations for constructing the international index numbers of agricultural production. The results provided an interesting comparison between countries of Africa, Asia and Latin America. The value of 'a kilogram of wheat equivalent' per person varies from 332 in Lebanon to 529 in Egypt and 1604 in Cuba.

So far as the present study is concerned all food items converted into calories and then spatial differentiation has been analysed between different districts of Bihar in terms of 'land resource measurement.'

THE CARRYING CAPACITY OF LAND

The carrying capacity of land expresses the resultant ratio between the productive capacity and environmental resistance. This should be measured in terms of food produce from the land either derived directly or indirectly. According to Marrak "one acre of land under potato would yield 3.7 million calories, wheat 1.8 million calories and by rearing cattle 0.16 million calories".[4] A calory is the unit of heat necessary to raise the temperature of one litre of water by one degree centigrade. This fuel energy consists mainly of fats and carbohydrates in the form of sugar and starch for human body. The calories also include a large number of items including carbohydrates, proteins and a large number of organic and inorganic substances which are necessary for maintaining the health and vigour of a human body. Thus, the carrying capacity determines the maximum number of people that a unit of area can support. The carrying capacity may improve with the increase in agricultural efficiency and vice versa. In Bihar per capita standard cultivable area is 0.49 acres which has resulted into deficiency of food and nutrition in different parts of the State. In the U.S.A. the Department of Agriculture (1934) arrived at the following estimates of the carrying capacity of land (land required per capita) for various levels and methods of food consumption and nutrition.[5]

"(a) An emergency restricted diet for an individual in any community on temporary basis consists largely of cereals and this will require—1.2 acres of land.

(b) For an adequate diet of minimum level with less cereals and more of milk, fruits, vegetables and meat—1.8 acres.

(c) For an adequate and moderately economical diet—2.3 acres.

(d) For a liberal diet meeting all the requirements of cereals and protective food like meat, milk, green vegetables, fruits and eggs—3.1 acres".

For estimating the availability of food for consumption it requires the consideration of addition of food by imports, deduction for wastage, loss by drying and for seeds, reduction of weight in course of preparation of grains and other food items. The entire produce of land is not consumed by human being but these are subject to various kinds of ravages and predations from wild animals, stray domestic cattle, birds, rodents and from thieves. After threshing commonly the grain is left in the sun-rays on the ground to dry and then that is kept in the storage bins. In this way the above processes of consumption of grains are automatically carried on every day.

In Bihar, there are three types of agricultural output e.g., grains, vegetables and fruits. Potatoes, oilseeds, sugarcane, gram and groundnut are also important in the scheme of crops grown in the region. However, gram, kodo, jowar and bajra are considered as meagre crops and constitute important items of diet for the poor.

In the State of Bihar, a large percentage of population is engaged in producing cereals which provide the bulk of the food for the peasants. On the other hand, only a small percentage of population supplements its daily food from such sources as egg, milk, vegetables, meat, fish and other eatables which are locally available in rural areas.

The measurement of land resource may be possible in terms of output per acre of land or per agricultural worker. In Bihar per capita cultivable land is less than 0.49 acre and hence per acre output of land is more important than the output per agricultural labourer. Nutritional value of food crops has been measured here in terms of calories and protein produce from grains, fruits, vegetables, etc. per oz. of weight. The calorific (C) and protein (P) value has been calculated with the help of Table 14.1

The concept of standard nutrition unit : The caloric intake is a good meaure of general health of a person. According to Prof. Stamp, the concept of potential productive unit (PPU) and standard nutrition unit (SNU) denotes—'the measure of an average yield of an acre of farm land of any area' and, the number of calories required for an average man' to keep him active and healthy respectively.[6] After considering a rigid economy, L.D. Stamp estimated that 1,000,000 calories a year must be produced in order to provide adequate food to each human being. He proposed to call it a 'Standard Nutrition Unit'[7] which might be related to the productivity of different crops on different types of land. This figure is consistent with the estimation made by James Wyllie[8] in the United Kingdom in 1936-50 who came to the same conclusion.

To measure the standard level of consumption, 'the Nutrition Advisory Committee, of the Indian Research Fund Institute' has laid down the following standard of a daily balanced diet for the maintenance of good health in terms of the constituents of Indian diet. The nutritional requirement is estimated below for an average adult engaged in sedentary occupation having a weight of 120 pounds (Table 14.2).

TABLE 14.1 : Food value of principal Indian types of food and vegetables (values per oz.)

Items	C	P
Almond	186	5.9
Apple	16	0.1
Bajra	102	3.3
Banana	43	0.4
Barley	95	3.3
Beef (Muscle)	32	6.4
Beet root	18	0.5
Bitter gourd	7	0.5
Brinjal	10	0.4
Broad Beans	17	1.3
Cabbage	9	0.5
Carrot	13	0.3
Cauliflower	11	1.0
Cheese	99	6.8
Chalam	101	3.0
Coconut	126	1.4
Cucumber	4	0.1
Curds	14	0.8
Dates	80	0.9
Drumstick	7	0.7
Duck	37	6.1
Egg (Duck)	51	3.8
Egg (Fowl)	49	3.8
Figs	21	0.4
Fowl	31	7.2
Garlic	40	1.8
Jack Tender	14	0.7
Jack Fruit	4	0.5
Ladies Finger	12	0.6
Lemon	16	0.3
Lentil	98	7.1
Lime	17	0.4
Liver (Sheep)	43	5.4
Mango (Green)	11	0.2
Mango (Ripe)	14	0.2
Milk (Cows)	18	0.9
Milk (Buffaloes)	33	1.2
Mutton (Muscle)	55	5.3
Onion (Largel)	14	0.3
Orange	14	0.3
Papaya (Ripe)	11	0.1

TABLE 14.1 (*Contd.*)

Item	C	P
Pear (Country)	13	0.1
Pears (English)	31	2.0
Pineapple	14	0.2
Pomegranate	18	0.5
Plaintain (Green)	19	0.4
Fomfret	22	5.4
Pork (Muscle)	32	5.3
Potato	28	0.5
Pumpkin	8	0.4
Radish	6	0.2
Ragi	98	2.0
Red Gram	95	6.3
Rice boiled	98	1.3
Grapes	9	0.2
Guava	19	0.4
Groundnut	156	7.6
Rice millch	98	1.8
Rice Homepound	99	2.0
Sago	100	2.4
Snake gourd	6	0.1
Spinach	9	0.5
Sweet Potato	37	0.3
Tapioca	45	0.2
Tomato (Green)	8	0.5
Tomato (Ripe)	6	0.3
Watermelon	5	0.1
Wheat flour	100	3.4

Where : C = Calories
P = Protein
1 oz = 28.350 g.
35.25 oz = 1 kg.
1 ton = 1016 kg,
1 tonne = 1000 kg.
1 oz. = ½ Chhatak.

Source : The Nutritive values of Indian Foods, I.C.M.R. No. 42, 1963.

TABLE 14.2 : **Nutrition requirement for an average adult**

Food Items	Dietary Allowances in oz.
Cereals	14
Pulses	3
Green leafy vegetables	4
Root vegetables	3
Other vegetables	3
Fruits	3
Milk	10
Sugar & Jaggery	2
Fat (Oil & Ghee)	2
Meat and Fish	3
Egg	1

Source: G.N. Sinha, *An Introduction to Food Economics*, Kitab Mahal, 1956, p. 70

The above table 14.2 shows that a sound body requires essential quantities of protective foods with mixed diet. The excessive reliance on rice or potatoes may give rise to serious illness. It is, thus, essential for a sound body to take mixed diet with varieties of food items.

TABLE 14.3: **Balance optimum daily diet**

Food items	Guha's (in oz)	Aykroyd's (in oz.)
Unmilled and lightly milled rice	10	10
Whole wheat	6	5
Pulses	4	3
Eggs	1 or 2	—
Sugar	2	—
Milk and Milk products	10	8
Fish and Meat	4	—
Non-leafy vegetables	5	6
Green leafy vegetables	5	4
Fats and Oil	2	2
Fruits	3	2
Per adult total calories required daily	2780	2600

Source : Mahto, R.B., Food Requirements in Bihar: A Critical Analysis, *Indian Geographical Studies*, R.B. No. 4, 1975, p. 68.

Similarly various attempts were made in the past to assess the daily requirement of balanced diet per adult. W.R. Aykroyd was inclined to lay down 2600 calories as the average daily requirement of a person in India. Aykroyd has planned a "well balanced diet"[9] and Guha has worked out a well balanced optimum daily diet as given in (Table 14.3).

This dietary allowance is for a sedentary worker and in comparison with this an average worker in fields and factories requires a higher level of nutrition. It is also a very liberal allowance for the tight economy of our country. On the basis of figures of our food supply whether produced or imported, one would be able to measure the calories and protein supply in our daily food for an average Bihari.

Similarly, the people of different age group require different amount calories per day. A growing child needs more calories than an aging adult. In addition to the calories, the food should also include adequate amount of protein of fruits, vegetables and animals. The Indian Council of Medical Research has carried out an exhaustive enquiry into various sources of food and has published a series of recommended dietary allowances of calories, proteins, calcium, iron, vitamin A & B and others for people of different age, sex and occupations. Thus according to the recommendation of the Nutrition Advisory Committee (NAC) of India the desirable amount of caloric requirement is 1790 a day for women and 2080 a day for a man in sedentary occupation and 2780 for persons engaged in light industrial occupation.[10] Among children it varies from 450 a day (after deduction of food received from mother) for infant under one year to 2675 for a teenage. The average for different categories comes to 2100 calories per head per day. The allowances of calories made by the Food and Agriculture Organisation (FAO) range from 1900 to 2700 for an average woman and man respectively.[11] The average in both the cases come to 2100 calories per day. Shafi[12] pointed out that there was an obvious evidence of malnutrition when the daily caloric intake falls below 2000 calories per day. In Table 14.4 is given a daily dietary allowance for healthy persons of different age groups:

TABLE 14.4 : **Daily dietary allowance for healthy person**

Age group	Required calories for male	Required calories for female
5—15	3000	2500
16—35	2900	2100
36—55	2600	1900
56+	2400	1700

Source : Food and Nutrition Board, National Research Council of U.S.A.

The average caloric requirement of a healthy person in age-group 18—35 is about 2500 calories per day which comes to a little over 900,000 calories per year. Stamp calls it as the 'Standard Nutrition Unit' (SNU). According to M. Shafi SNU for an Indian farmer it should be fixed at 800,000 calories a year, because Indians generally require less calories than an American. This is also supported by the recent work of Canadian army that generally people living in cold countries require more calories to maintain good health than those living in hot climates.

In assessing the total amount of SNU produced in the State of Bihar in 1971, the extraction rate for various food items has been derived from Table 14.1. It is also essential to reduce 10 per cent to 20 per cent of food produce for seeds. Further, in any planning of land management land needed for non-food crops should be given additional preference.

TABLE 14.5: **Production of calories and protein in the state of Bihar, 1971 (in crores)**

Food Items	Calories	Protein
Goats meat	7,580	1663
Buffaloes milk	45,550	1656
Cow's milk	20,247	1017
Potatos	6,75,324	1206
Other vegetables	14,904	447
Other Fruits	61,819	2013
Peas	86,063	1555

Fowls Meat	606	4
Eggs	44,488	3381
Sugarcane	13,60,023	4405869
Arhar	41,162	840
Other Food grains and pulses	2,95,559	13669
Rice	14,49,742	29288
Wheat	4,43,768	15088
Maize	3,72,798	7848
Ragi	8,400	644
Jowar-Bajra	31,191	183
Gram	58,346	3869
Barley	54,737	1901
Bihar Total	50,72,309	4490442

Source: Season and Crop Report, Government of Bihar 1971 and Nutritive values of Indian Food ICMR, No. 42, 1963.

According to Table 14.5 it is apparent that at 800,000 calories per person consumption in a year, Bihar can support only 3.44 crores of people with food grains produce and 6.54 crores of people with its vegetables, fruits, milk and other eatables. S N U for total calories produce comes to 887904 per person in a year, while it comes to 489468 for food grains produced in the State of Bihar. But it is a sad state of affairs that more than three-fourth produce of sugarcane is sold outside the state and a large amount of potato, dairy produce and fish are sold in the nearby cities and towns and Calcutta conurbation. The fruits are grown seasonally and one abundantly available only during June-July. In relation to the real world situation, it is apparent that the state is not able to support its 5.63 crores of people properly as the problems of flood and drought quite often make their way over most parts of Bihar. Mostly after each flood and drought malnourishment prevails in Bihar and as a consequence of this a large number of people migrate periodically to Calcutta conurbation from each and every rural settlement.

At present the gap between the average SNU of 800,000 calories as fixed for the average Indian adult and SNU produced by food grains produce of only 4,89,468 calories in Bihar is due to the lesser production of foodgrains in comparison with the population pressure of the region and hence the rest is compensated by other means. Thus the further increase in population requires more efficient use of the land. This ensuing problem invites us to consider the introduction of scientific cultivation of land and the best cropping pattern in the State of Bihar besides raising considerable amount of fruits, vegetables, fishes and several other items of our daily food.

The use of Standard Nutrition Unit

1. In comparison with the SNU, the production and consumption of food crops should give the deficit and surplus areas of land resource.
2. Total food output and population pressure of the region may give the farming conditions of the region in relation to crops grown.
3. Agricultural efficiency may be measured with the help of SNU obtained in a particular region.
4. Population supporting capacity or carrying capacity of the land may also be measured.
5. The productive capacity of different types of land i.e., the range of land capability classes can be measured.

General problems and prospects : The most important problem which our country is facing today is the shortage of food for our fast increasing population. In spite of the best efforts, we are still importing a considerable amount of food grains from various parts of the world. Imports can offer only a temporary solution of the problem and hence our home production should have to be increased to feed our people.

The nutritional value of land has been measured here in terms of its use, calories and protein produce in different districts. In Bihar, the northern part is fertile plain but the havoc of flood each and every year, irregular monsoon rainfall, lack of irrigation facilities and high population density make the inhabitants poor in most cases.

In recent years with the attention of the Government towards assuring irrigation facilities, Bihar is now approaching towards self-sufficiency in food produce. In this regard the public and the private irrigation management have helped much with the use of need based chemical and compost fertilizers. These have also raised the general productivity of the region.

On the other hand, in Chota Nagpur Plateau rugged forest clad terrain restricts the high density of population, as stony wastes are unsuitable for raising good crops. In spite of this, Chota Nagpur is rich in mineral deposits which have raised the levels of cultural advancement and the people have now greater incentive for scientific cultivation of fields. In recent years Chota Nagpur also produces high yielding varieties of rice, wheat and fruits with the help of chemical fertilizers.

The above discussion clearly reveals that the State of Bihar is in a position to achieve self sufficiency in food, if the government come forward with need based capital investment and irrigation.

REFERENCES

1. M. Shafi, Can India support five times her population? *Scienc Today*, May, 1969, pp. 22-23.
2. Ibid.
3. J.L. Buck, *Land Utilization in China*, 1937.
4. G.N. Sinha, *An Introduction to Food Economics*, Bihar University Publication, Allahabad, 1956, p. 34.
5. U.S. Department of Agriculture, 1934, Quoted by G.N. Sinha, *An Introduction to Food Economics*, Bihar University, Patna, 1956, p. 34.
6. V.R. Sinha, Carrying capacity of land around Mirzapur town, *The National Geographical Journal of India*, Vol. X, pt. I; 1964, pp. 29-37.
7. L.D. Stamp, The measurement of land resources, *The Geographical Review*, Vol. XLVIII, 1958, No. 1, pp. 1-15.
8. J. Wyllie, *Land Requirments for the Production of Human Food*, London, Wye College, 1954.
9. W.R. Aykroyd, The nutrition value of Indian foods and the planning of satisfactory diet, *Health Bulletin*, No. 23, 1962, p. 21.
10. L.D. Stamp, op. cit, 1958.
11. Ibid.
12. M. Shafi, *Land Utilization in Eastern Uttar Pradesh*, Muslim University, Aligarh, 1960, p. 222.

15

Conservation of Land Resources

LAND is the basic resource of human society. The word conservation means the preservation of natural resources e.g., soil, vegetation, water and animals from destruction, by careful control and management, especially for the benefit of the coming generations. This does not mean a 'holding-back' for the maintenance of favourable balance in the use of the environment.

In the assessment of land resources for conservation, the land use study has assumed a place of pride in the field of applied geography. It has a special significance in a predominantly rural areas where cultivation has been carried on from times immemorial and where agriculture is almost the sole support of the increasing population pressure on land.

The pioneer of conservation movement was Gifford Pinchot of the U.S.A. In 1907 Taft defines conservation as 'the use of natural resources for the greatest good, to the greatest number and that for the longest time'. But in practice, the maximization of the three superlatives i.e., greatest good, greatest number and the longest time as used in the definition may not be possible at the same time.

Thus, the concept of conservation means that the resources should be used for the maximum benefit of a man both at present and in future. The production potential of the use of land must be geared up to meet the demands of consumers, in terms of kinds, amount and the qualities of goods and services forthcoming from our land. The cost involved should be as low as the technical abilities permit and the productivity of the land supports.

The use and conservation of land and water takes place within three dimensions i.e., physical, economic and legal.

Physical dimension : The physical dimension deals with kinds of plants, yield of crops, yield of specific crop with different methods of cultivation, irrigation, application of fertilizer and the aggregate rate of input. This also includes the contour interval, terraced cultivation, water retention periodicity and their effects on soil erosion, silting, crop yields, etc. The physical dimension of the conservation is being extended constantly through the technological advancement and investigation.

Economic dimension : This includes in land, prices of products, and the relationship between products, labour and capital in harnesting the resources of the land. The economic possibilities of conservation are constantly changing with the fixation of price, cost-benefit control and the demands of consumers.

Legal dimension : The legal dimension consists of a set of rules and customs that were made by the man. It can be changed in different situations to meet the requirements of man and the conservation of land and water resources. This may be clearly expressed i.e. a tank may supply 10 centimetres of water for 1000 hectares of land under rice crop. This shows the physical possibility of the tank and availability of water, but the cost of supplying water in relation to price of products may mean that only 5 centimetres of water can be used profitably under economic possibilities. On the other hand, the water may also be restricted for the purpose of irrigation and then in that case, it would not be possible under legal dimension to use water for irrigation purposes.

CLASSIFICATION OF RESOURCES

The aims and objectives of conservation differ with varying land resources. This necessitates classification which is based on the relative renewability of resources. The classification is as follows: (1) Fund resources, (2) Flow resources and (3) Composite group of resources which have some characteristics of both fund and flow resources. The composite group is again sub-divided into (a) Biological resources (b) Soil resources and (c) Man made resources.

Fund resources : The resources, as minerals, metals, petroleum and coal are classified as fund resources. The physical supply of these resources is limited over landscape and hence these cannot be replenishable. The minerals like coal and petroleum are exhaustible, while metals and stones wear away slowly and are often capable of renewed use. This in conservation, fund resources call for reduced rate of consumption and increase in surplus supply for the future use.

Flow resources : The rainfall, river water, tides and wind are flow resources in the form of stream and weather, whether we use it or not. These resources should be used for the benefit of mankind, and the failure to use them results in their permanent loss. Flow resources may be captured and stored for future use, and in that sense a reservoir full of water may be called fund resources. Flow resources require maximum utilization to avoid unnecessary wastage.

Composite resources : Conservation of biological, soil and man-made resources calls for highest possible return at the cost of maintaining and improving their expected productive capacity.

Biological resources : This includes crops, forests, pasture, livestock, wild life, fish and even human beings. These resources can be replaceable over time, if planned carefully for regeneration. At any given time, however, these can be treated as fund resources. The productivity of biological resources may be increased or decreased or maintained at constant level, like fund and flow resources.

Soil resources : This represents a combination of fund, flow and biological resources. The exploitation and maximum use of the fund of fertility stored over a longer time is dependent upon the farmer. On the other hand a farmer may build up the productive capacity of the soil. Soil lacks the characteristics of life cycle unlike plants but their productivity can be increased or decreased, with access and low land use practices.

Man-made resources : This includes all the cultural features e.g., settlements, houses, multi-purpose dams, industries, humdrum reality of urban life and all these depend upon man to understand the economic usefulness and raise the life expectancy of resources. A house requires upkeep, repairing and occasional remodelling to aid new features. In absence of these, the durability of a building may be adversely affected by abuse. Hence, with the help of timely repair and good management the longevity of a house or any resource can be enhanced. It is, therefore, rightly said that—'A stitch in time saves nine'.

Most of the man-made resources are the results of human knowledge. The coal is found in nature but it can never be treated as a resource, unless it is investigated and extracted from the earth surface. Unlike this, there are several items, e.g. coking coal, ammonium sulphate, nylon, paints, medicine etc. which are invented by the ingenuity of man. All the minerals, elements and compounds are available in nature but these can be utilized as resources only after their proper knowledge and extraction from the nature. Thus, greatest among the human resources is the knowledge. The concept of resources is functional and it is inseparable from the requirements of man and human efforts. Man is also an originator and destroyer of resources, originator in the sense that each and every resource is the product of human knowledge and destroyer in the sense that once coal is burnt, the same cannot be replaceable for ever.

PROBLEMS OF CONSERVATION

There are fourfold problems of land conservation i.e. physical, economic, institutional and technical. Under physical obstacles, the changing nature of the earth's surface and the vagaries of nature put hindrances in the proper utilization and conservation of land resources. Only after proper investigation, one could be able to locate and dig mines, build dams, tame the mighty rivers, construct roads and bridges and use terraces to raise crops. Thus mainly topography of the land and climatic barriers at some places as on the equatorial belt especially forest covered mountainous terrain create problems of land resource conservation.

Economic obstacles include the lack of capital and the knowledge about the capacity of resources to give maximum profit. The economic instability can easily be checked by controlling the price of commodities in the market and stabilizing the economic system.

Institutional problems include conservation decision by institutional factors of habit, customs, land tenure rights and insurance principle. Under this head, the government should give a subsidy to the education to boost up the knowledge of conservation practices.

The hindrance of technical abilities can easily be removed by organizing the training camps and establishing resource based industries etc. to conserve the land resources. Not to use technical knowledge for the sake of conservation is an unscientific approach and that should always be discredited.

PROBLEMS OF LAND CONSERVATION: A CASE STUDY OF INDIA

The spatial variation in land use is the product of varying ecological conditions in which the conservation needed to check sheet erosion of soil layers, especially during floods and heavy monsoonal rains, takes place. This takes place in the plains and mountainous slopes. The soil erosion can easily be watched as to how soil wear away from fields by the monsoon rains. As for example, in the blacksoil fields of Sholapur district in Maharashtra, soil, on average of 328 tons per hectare is lost per year. This is responsible for a greater loss of production than the aggregate forms of other shorts of erosion, such as with gully erosion, lands in the adjoining areas of banks of rivers and channels are cut up into ravines for long distances in the Chambal valley. Gullied land in Uttar Pradesh, Punjab and North West Frontier Province of Pakistan amounts to about 8000 square kilometres,[1] and these lands are not only rendered completely useless in themselves, but they also pose a serious danger to the neighbouring good land. Similarly, vast expanses of deep alluvial soil of exceptional productivity in northern India converted into useless badland topography in the Bundelkhand and Malwa plateau on which only a few goats and cows can be reared miserably. Still further, bank erosion occurs, to some extent, along the channels of all Indian rivers and leads to a serious loss of cultivated land, when large slabs of riverside land are broken off and swept away. These are the problems of conservation of natural resources and better use of the land. The unwarranted cutting of forests provides free movement to flood water, and this brings vast amount of micaceous sand from hill slopes. In this way, the river beds are choked up and flood water spread over adjoining areas, causing extensive damage to cultivated area. The continued process of this enormous silting may hamper the life of dams and reservoirs, if this state of things prolongs unchecked.

Only afforestation programme can help in checking the soil erosion. In some areas, the destruction of forests is due to the wasteful methods of shifting cultivation, as practised by the inhabitants in the upper reaches of the plains which are largely tribal areas. The Government of India has adopted programmes of irrigation and afforestation in the catchment areas of rivers and river valleys to conserve our land resources. The facts show that the practice of soil conservation should be adopted to raise agricultural productity and soil fertility. Thus, the programme of conservation should be made in a coordinated manner for the region as whole. The measure of conservation of land fall into three classes in rural areas, which are as follows:[2]

(a) Those which have to be undertaken on cultivator's holdings e.g., contour bunding, gully plugging and strip cropping etc.
(b) Those which require collective efforts in the village as a whole-control of erosion in the catchment within the village limits of small streams control of grazing in the village common land or raising a fuel or fodder reserve for a village.
(c) Common works for groups of villages in the region, incorporating the catchment of a stream or streams.

Land is the best resource in the present day world where this civilization stands and supports millions of people at a time. Hence the misuse of land means the loss of resources and its conservation and the judicious use means greater good of human society.

REFERENCES

1. R.M. Gorrie, National Planning Committee Report, April, 1956.
2. V.T. Krishnamachari, *Planning in India*, Orient Longmans, Calcutta, 1961, p. 177.

16

Land Reforms in India

LAND REFORM is necessary for the emancipation of peasantry from the economic ills. It should be a part of the process of national planning for economic development. The historical perspective of land reform referred that it was the Russian Revolution which provided the necessary ideological inspiration for land reforms to the nationalist fighters in the colonies or semi-colonies.[1] In modern period, land reforms came to be recognized as a fundamental condition for economic development of the newly liberated countries.[2] It is, thus, rightly felt that land reform is an enormously important issue for theoretical discussion and empirical investigation. It has been highly emphasized in the reform of agrarian structure by Doreen Warriner in her lectures on 'Land Reform and Economic Development'.[3]

MEANING AND SCOPE

Land reforms mainly include the consolidation, distribution of surplus land received from ceilings abolition of intermediaries, reduction of rent, resumption of land for personal cultivation, voluntary surrender of surplus land and preservation of the ownership of tenants, etc. Unfortunately, in India the success of land reforms is meagre and the government have failed even in preventing sub-letting and rack-renting and what to talk about the restoration of right to cultivate land to the tillers of the soil.

The term 'land reforms' has been analysed in various ways due to varying objectives in adopting this measure. However, the establishment of direct link between the tenants and the State Government, and the abolition of the rights of *zamindari* paved the way for further land reforms, both in the narrower as well as broader sense of the term. These measures are intended to serve two purposes viz: (a) social justice and (b) improvement in agricultural production. Considering the historical perspective Indian National Congress has always been attaching great value to the question of achieving for reaching land reforms in India by eliminating the feudal intermediaries and by bestowing security of tenure and economic stability on the peasantry.

The 'World Land Conference (1966)' defined land reforms as "a hotch-potch of measures designed to eleminate obstacles to economic and political development arising out of the defects in the agrarian structure". In India it is formulated in terms of the enaction and implementation of law pertaining to the ownership and occupation of land for the efficient cultivation and raising bumper crops.

Land is a resource touching every aspect of human living. It runs like a thread through almost the whole fabric of geographic and economic thought. It has, thus, intimate relationship with all the resources.

Hence, land reforms and planning precepts for boosting up agricultural production are twin aspects of agrarian society which should be discussed among the scientists of all the disciplines.

The tillers of the land are poor labourers and share-croppers, while the excess land owners are always rich persons. It is for this fact that the agricultural labourers have neither the means nor they have any interest in taking adequate measures for improving agricultural productivity. But now, with the changing time, increasing population density, restriction on land occupancy, enhanced land values, scarce cultivable land and problems of food shortage necessitate land reforms to meet the present requirements of the society. Land reforms is, thus, the creation of a social-order relating to the ownership, occupation and equitable distribution of land among poor peasants and landless agricultural labourers, so that the efficiency of agricultural production may be raised on the ground of social justice and economic prosperity.

Land reforms involve the proposals of acquiring surplus land and its redistribution among the landless. But there is also the other side of the coin. Suppose we have no land at all. In such circumstances, the existing land cannot be expanded like rubber, rather we have to think over for the better utilization of the available land.

In any country, the form of government and the society may be either capitalistic or socialistic but the problem of land reforms overwhelms all of them. Regarding land reforms these questions are often put forward.

(a) How can land fulfil utmost needs of the society ?
(b) What should be the pattern of land use ?
(c) What sort of land management is necessary ? and
(d) How the measures of land reforms should be implemented to achieve the target ?

Land reforms are often considered in a narrow sense of the division and distribution of land to the people of lower income group and the legal laws relating to the land. But the land reform should aim at the maximum benefit of the society. The farmers and the farm labourers should have equal share of the land. Generally three types of land ownership have been suggested.

(a) The land should be in the hands of a handful of persons and it should not be divided as the application of new scientific implements is not possible on smaller farms.

(b) All persons should be given equal share of the land but the same is not possible in the present set up of our society, as the nature knows no equality and identity. Thus, whatever may be the intensity of reforms galore the 100 per cent land reforms cannot be achieved.

(c) The small farmers produce much with hard labour hence the land should be given in the hands of the poor.

At this stage, it can safely be said that the removal of hurdles of the use of scientific implements may easily be effected with the cooperative farming on consolidated farms and this should be preferred for the success of land reforms programmes and the better use of the land.

WHY STRESS ON LAND REFORMS

India is predominantly an agricultural country where 68 to

69 per cent people depend on land for their livelihood in comparison with 6 per cent in the U.S.A., 20 per cent in the U.K., 50 per cent in France and Germany. Land reforms have two aspects : social justice and equality before the law. In case one's economy is not planned then social justice necessitates that each and every person of the society must have an equal share in the land so that the bare necessities of life may be fulfilled. So far as the equality before the law is concerned, the justice demands that as the land is a natural resource, every individual should have equal right to it like air and water. Keeping these ideas in mind Vinoba Bhave started 'Bhoodan Gramdan Movement' in 1951 in which he has obtained more than 23 lakh acres of land up to 1970 in India.

The idea of land reform is also stressed by Gunar Myrdal in his celebrated book *Asian Drama*. He asserted the need of purposeful reorganization of ownership and tenancy of land with the help of credit reorganisation, agricultural education, provision of improved high yielding seeds, pesticides, insecticides, irrigation facilities, proper marketing of produce, advisory service to improve the conditions of Indian peasantry and better use of the land.

The Indian society is built on the basis of equality but owing to lack of opportunity in obtaining facilities on land, all are different from one another. Nature does not know identity and hence the equal distribution of land to all is a less possibility, but even the needful accessory to good life must be granted according to our constitutional law.

The exact problem in land reform is how the tillers of the soil may get the land from rich persons, as the poor have no land at all. This is because the people have ever been considered the constraints of economic development and the progress of land resources, while land is a more important factor than all other factors of production.

In India, the majority of land holdings are economically not viable, the production is low and the man-land ratio is high. A well planned measure of land reform can act as a catalyst that sets in motion the whole system of economic, social and political changes of land use pattern and agrarian social make-up for economic development.

There are two views on the concept of land reforms, i.e.,

the traditional and the modern. The traditional and most accepted view is the social justice and the equitable distribution of land with the transfer of ownership to the tillers of the soil. Undoubtedly, this may lead to tenurial changes and imposition of ceilings on agricultural land holdings in rural areas.

According to the modern view land reform should include redistribution of land and the covering of development oriented package programme. This requires a set of policies for rational use of the land, consolidation of holdings for efficient cultivation of land, tenancy reforms, improvement of credit system, cooperative organization, agricultural training, etc. In this way the report of Food and Agriculture Organization has rightly emphasised the role of land system in influencing the rate of agricultural development. "Unsatisfactory form or condition of land tenure may constitute a major impediment to development by creating or perpetuating social unrest as well as by hampering the modernization of agriculture. Out of date cropping system, for instance may be dropped by tenancy laws and customs. The application of modern methods may be impossible, because the farmer's income are depressed by exorbitant rate to an extent which leaves no margin for investment. The tenant may lack minimum security of tenure which would encourage him to invest savings in improvement".[4] Agricultural land use from all over India shows that faulty system of tenancy reduces output. "Land reforms, therefore, are not just a matter of social justice to the peasants but also vital for providing suitable institutional framework for agricultural development. Hence the measures of land reform occupy pivotal position in the planned development of agricultural land use".[5] Through this study an assessment has been made about the land reforms in India and a progressive view has been suggested to weed out the hazards of land tenurial system for efficient cultivation and raising better crops. This is urgently needed to support the increasing population.

FACTORS AFFECTING LAND REFORMS

The land is a natural gift to mankind. Only 42.4 per cent land in India is under cultivation and the rest are forested, rugged terrain, riverine diaras and flood affected marshy ground. And hence, land reform is applicable only to those lands which are cultivated by men. No one is desirous of reclaiming forest margins and marshy ground but whenever some one reclaims it, the measures of land reforms are at once applicable even to those lands. In this way the factors which affect land reforms are mainly cultural factors which are as follows :

Parameters Affecting Land Reforms

- History of land occupancy
- Population pressure
- Population growth
- Percentage of agricultural labourers
- The will of the people for such a change
- Government participation
- Political system
- Land available for cultivation
- Land laws

India has an age long history of land occupancy. It has seen several stages of settlement growth, human migrations and administrative reforms from Prehistoric period to the Modern period. Thus the history of land occupancy is so eloquent that it is difficult to change the traditional ownership of the land in a day.

In India the population pressure in rural areas is higher in comparison with that in the other developing countries of the world, which requires urgent attention to land reforms. In the Himalayan Foothill zones of India 100 per cent population is rural. This ratio is more than 90 per cent in plain lands, while in the mineralized belts of Chota Nagpur Plateau, Madhya Pradesh and Maharashtra, more than 80 per cent people depend for their livelihood directly or indirectly on land.

The population growth rate is very high in India. The growth rate of total population reveals that the average percentage rate is 24.80 per cent (1961-71). In such a situation the redistribution of the existing cultivable land is amply justified, which is concentrated in a handful or persons of the society.

According to 1961 Census the incidence of agricultural labourer was 23.65 per cent for India as a whole, while in 1971 Census the percentage of share-croppers and landless labourers was 25.8 per cent for the whole country. These figures are indicative of the fact that the early implementation of land reform measures are necessary but the success is possible only when people in general have will for such a change along with the governmental participation, but the question arises as to how much land is available for cultivation. In case the population pressure is high and the cultivable land is scarce then what will be the use of distributing just a small piece of land. Further, the land laws are the limiting factors of land reforms measures, because each society and the government have legal bonds which cannot be broken without the consent of the two. Thus the carriers and barriers of land reform measures are the social order, and they care for the govermental participation to effect any change.

SYSTEMS OF LAND REFORMS

The intensity of land reforms may be explained in a systematic order from ancient to modern period through a dynamic land reform model (Fig. 16.1). In ancient period the land reform was almost non-existent and the land was free to all as the population was sparsely distributed. In medieval period with the increasing pressure of population, the intensity of land reform increased to medium level as the land was less free in comparison with that in ancient period. So far as the modern period is concerned, with the higher population pressure, the land is not free and the need for the measures of land reform is intensely felt in a traditional Indian culture where the lust for landed property is deep-rooted as the 68 per cent population of our society are directly deriving their sustenance from the land.

POLICIES ON LAND REFORMS

Seven point policies have been adopted for land reforms in India, which are as follows :

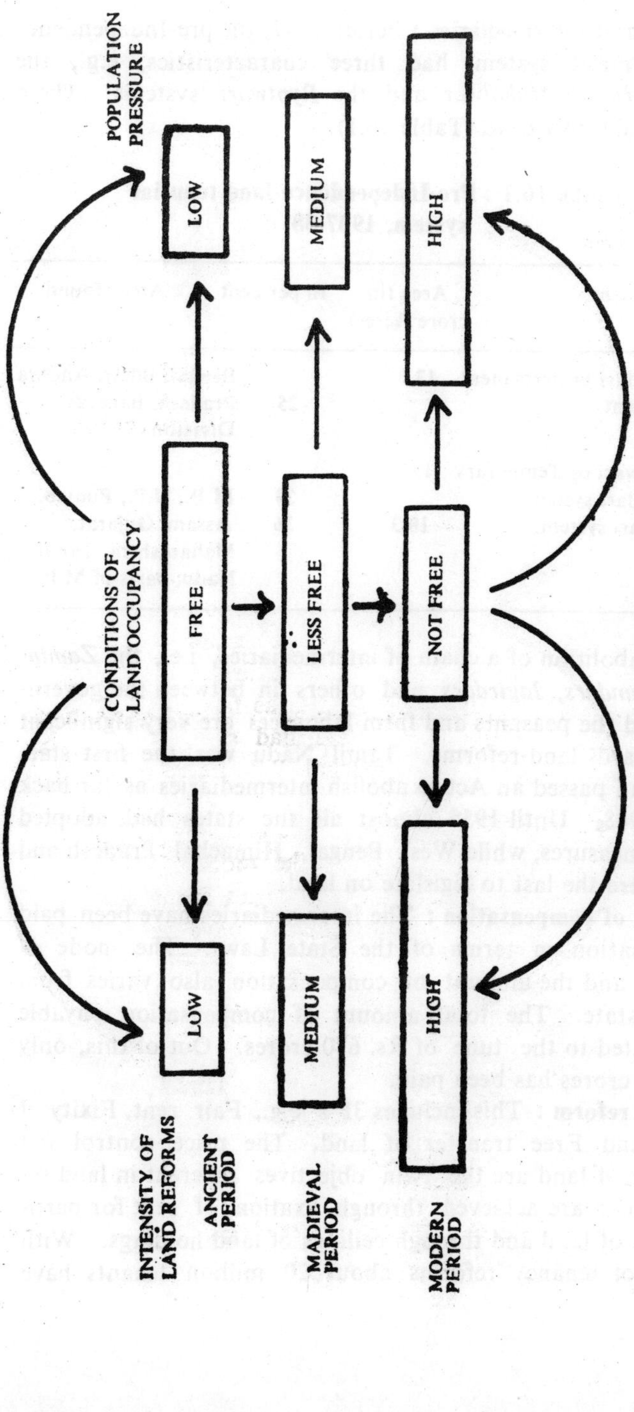

FIG. 16.1 : *Dynamic of land reform model*

Abolition of intermediaries : Before 1947, the pre-Independence land tenurial system had three characteristics, e.g., the *Zamindari*, the *Mahalwari* and the *Ryotwari* systems. These systems are now dead (Table 16.1).

TABLE 16.1 : **Pre-Independence land tenurial system, 1937-38**

System	Area (in crore Acres)	In per cent	Areas found
1. Zamindari or permanent settlement	13.0	25	Bengal, Bihar, Andhra Pradesh, Banaras Division of U.P.
2. Mahalwari or Temporary Zamindari system	19.9	39	M.P., U.P., Punjab.
3. Ryotwari system.	18.3	36	Assam, Gujarat, Maharashtra, Tamil Nadu, parts of M.P.

The abolition of a chain of intermediaries, i.e., the *Zamindars, Inamdars, Jagirdars* and others in between the government and the peasants and farm labourers are very significant step towards land-reforms. Tamil Nadu was the first state which had passed an Act to abolish intermediaries as far back as in 1948. Until 1955 almost all the states had adopted similar measures, while West Bengal, Himachal Pradesh and Delhi were the last to legislate on land.

Payment of compensation : The intermediaries have been paid compensation in terms of the State Laws. The mode of payment and the amount of compensation also varies from state to state. The total amount of compensation payable is estimated to the tune of Rs. 670 crores. Out of this, only Rs. 2.75 crores has been paid.

Tenancy reform : This includes 3F's e.g., Fair rent, Fixity of tenure and Free transfer of land. The price control and rationing of land are the twin objectives inherent in land reforms which are achieved through fixation of rent for particular use of land and through ceilings of land holdings. With the help of tenancy reforms about 20 million tenants have been vested with proprietary rights and brought into direct contact with the State Governments in India. This has decreased the rate of eviction to a considerable extent, and the land reforms have made the tenants conscious of their rights. The tenant and the share-croppers with insecure tenure are estimated to constitute 82 per cent of the total number of tenants mainly in the States of Tamil Nadu, Bihar, Haryana, Punjab, Assam, West Bengal and Andhra Pradesh. The insecurity of tenancy has not only impeded the widespread adoption of high yielding variety seeds, but in some cases it has even led to social tensions and agrarian problems.

The national guidelines issued by the Planning Commission in regard to tenancy reforms include the following main points :[6]

(a) Maximum rent should be fixed at one-fourth to one-fifth of the gross produce.

(b) Issuance of receipts for the rent collected should be made obligatory by providing deterrent punishment in case of default.

(c) All tenancies should be declared non-resumable and permanent.

(d) Sub-letting of land in future should be prohibited except where the land holder is serving in the armed forces or is suffering from a specified disability.

(e) Personal cultivation should be so defined as to really mean self-cultivation.

Ceiling of land holdings : In several states ceiling laws has been invoked on land holdings. Bihar government passed the ceiling laws in 1961; Tamil Nadu and Kerala in 1970 ; West Bengal in 1971 ; Assam and Haryana in 1972 ; Himachal Pradesh, Orissa, Punjab, Rajasthan and Uttar Pradesh in 1973 ; Gujarat, Karnataka, Madhya Pradesh and Tripura in 1974 and Andhra Pradesh passed the Act in 1975.

So far as the implementation of the Ceiling Act on land holdings are concerned, this involves four stages : first, legislation imposing ceiling ; second, declaration of surplus land ; third, take over of the surplus land by the State Governments and fourth, the distribution of surplus land. But the Land Ceiling Act has not yet been fully implemented.

In different states of India, keeping in view the sociopolitical conditions have fixed the ceiling on land holdings such as 10 to 60 acres in Bihar, 27 to 132 acres in Gujarat, 22 to 336 acres in Rajasthan, 76 to 126 acres in Maharashtra, 20 to 80 acres in Orissa, 40 to 80 acres in U.P., 30 to 80 acres in Punjab and Haryana and it was 12 to 15 acres in Kerala.

In 1972, All India Congress Working Committee (AICWC) laid down uniform levels of land holdings which has given six clauses as guidelines to the different states for its implementation: (i) The land with assured irrigation and capable of yielding at least two crops a year should have a ceiling of 10 to 18 acres, (ii) the land having assured irrigation and only one crop yield, the ceiling could be raised to 27 acres, (iii) on all other types of land the ceiling should be 54 acres, (iv) in case of desert and hilly areas, the limit is also that of 54 acres, (v) the family should be considered as an unit of ceiling, (vi) the State Government must pass a legislation to this effect up to December 31, 1972 and its implementation with effect from June 30, 1976.

The next step is the 'declaration of surplus land' and the distribution of that 'surrendered land' among the tenants. Till December, 1975 only 5.88 lakhs acres have been declared surplus, out of which 2.92 lakh acres have been taken and distributed only 1.52 lakh acres to 54,000 persons only.

Settlement of agricultural labourers : The settlement of agricultural landless labourers is a challenge to the land reform protagonists and the solution lies in the regulation of rent and security of tenure. Rack-renting which is still prevalent in some quarters if not checked may serve as an impediment to investment. Later on it was recommended that the rent should be paid in cash and receipts should be issued by the landlords and later on by the government. Now some states have prohibited the creation of tenancy or the cultivators have been given right to enable them to obtain institutional credit. In some states agricultural labourers have been given land from the surplus land along with some base money as input for producing crops. The voluntary organizations, cooperative and rural banks and the State Governments are giving the institutional credit to the agricultural labourers. Under 20-Point economic programme, in view of providing house sites to

.hese people, the Government of India has sanctioned a grant ɔf Rs. 108 millions in 1976.

Consolidation of holdings : The sub-division and fragmentation ɔf holdings are great hindrance to Indian agriculture. In land reform measures, the consolidation of holdings is a good step n order to boost up agricultural production. Since the last generation, the campaign for the consolidation of land has been launched but the compulsory legislations to this effect have been passed only in Bihar, Gujarat, Madhya Pradesh and West Bengal. This work has been completed in the Punjab, Haryana, Uttar Pradesh and some headway has been made in Maharashtra and Himachal Pradesh (Table 16.2)

TABLE 16.2: **Progress of consolidation of holdings in various States of India: 1960-74 in '000 hectares**

States	1960-61	1973-74 (estimated)
Andhra Pradesh	127	337
Assam	—	43
Bihar	24	236
Gujarat	375	1258
Haryana	—	3273
Jammu & Kashmir	—	65
Madhya Pradesh	1560	4249
Maharashtra	626	7077
Karnataka	406	1322
Punjab	6011	9203
Rajasthan	571	1744
Uttar Pradesh	2201	12281
West Bengal	—	710
Delhi	83	83
Himachal Pradesh	66	267

Source : India: A Reference Annual 1975, Government of India, Publication, New Delhi.

India is a land of small holdings where the average size of farm is of 2.5 hectares. In total, 20 per cent of the holdings are below 0.46 hectares. It is felt that consolidation work has no tangible effects to solve the problem of small holdings but it will certainly help in the economic operation on the existing holdings. Moreover, small holdings can be pooled for the purpose of cooperative farming.

Prevention of fragmentation : Effective steps must be taken to prevent the fragmentation of land holdings. Some states have passed laws prohibiting the division and fragmentation as it leads to uneconomic size of holdings and unprofitable cultivation. Cooperative farming is an effective step to check it but the absence of institutional incentive is a great hindrance to its successful implementation.

TABLE 16.3: **Estimated number of fragments per operational holdings in India**

Size of group (hectares)	No. of fragments per holding	Average area of fragments (hectares)
Upto 0.20	1.61	0.06
0.20 to 0.40	2.82	0.10
0.40 to 1.01	4.41	0.15
1.01 to 2.02	6.30	0.23

Source: *The Gazetteer of India*, Vol. III, Government of India, New Delhi, 1975.

PROGRESS IN LAND REFORMS

The implementation of land reforms in India is not only a political necessity but it has also been accepted as the *sine qua non* for economic restructuring and social regeneration. The 20-Point economic programme is a safety valve for the Fifth-plan which consists of the following points regarding land reforms.

Institutional changes : The plan recommends the amendment of defective tenancy laws to ensure security of tenure and proprietary rights to the poor farmer. The present practice of 'personal cultivation' should be curbed to reduce eviction of tenants. Absentee ownership and the presence of benami land ownership must be abolished as the State Government are still inactive to trace out the land owned by the same person in different villages and districts.

Operational programme : This programme includes (a) preparation and maintenance of tenancy records, (b) uptodating of the records of tenancy rights, (c) cadastral surveys of the tribal areas, (d) confirmation of statutory powers on State Government for undertaking work on (a) and (b), (e) distribution of surplus land, (f) timely support of inputs to help farmers to bring land under cultivation and (g) consolidation of holdings operations.

Implementation machinery : For the successful implementation of land reforms, the Planning Commission felt the need for a new administrative set-up which will be hierarchical in structure, having a land reform Commissioner at the apex and land reform officers at the district, tehsil and village levels. It was also felt that the village level functionaries like, the *patwaris*, the *karamcharies*, etc. have been seriously hampering the implementation of land reforms. Hence, hand-picked officers should entrust this task to the one who may face any challenge boldly.

Land reform tribunals : It was realised that land reform measures be placed beyond review by the courts. The plan envisages the constitution of land reform tribunals which will function like itinerant courts to administer justice to the agricultural labourers, the landless poor or the share-croppers.

People's participation : It is suggested in the plan that at block and village level, an organization of peasants should be formed by election of 5 to 7 members from the group of landless labourers, share-croppers and small land holders of the village who owned less than 0.5 hectares of land. From each village one member should be elected for the block level committee.

Financing of land reforms : There is a provision of 151.85 crores of Rupees in the Fifth-plan for investment in various programmes of land reforms. This fund is meant for the assignees of the land consolidation of holdings and preparation of tenancy records, etc. In earlier plans there were no provision for financial assistance.

Mass campaign of land reforms : A sound public assistance of great help in advancing and implementing land reform programmes. Hence the mass media, the press, the radio and the television may be pressed into service. This can also be possible with the assistance of Gram Panchayats, Block Development Committees and District Development Committees and District Development Councils. The dissemination of information on land reforms is essential in view of the surging needs of the society and economic backwardness.

Till March, 1976, 18 lakh acres of land have been declared surplus in India, and this has to be distributed among 67 lakh landless labourers. This means that one family will get a small piece of land which is not economically viable. In the U.S.A. on an average three persons owned a motor car but in India even the food is not available to a large group of persons daily. Thus in India food problem and economic backwardness should be solved in close association with the land reforms and the optimum utilization of agricultural land.

In India, the administrative machinery of State Governments has also geared up its activities in settlement of government land. According to the present policy, members of the scheduled castes, scheduled tribes, backward class and army personnel are given priority in settlement of vacant government land in rural areas for agricultural purposes.

The problems relating to the land reforms are more practical than others. The main factors for the tardy progress in implementation of land reforms are the State Governments themselves who have failed so far in generating political will, firm determination, better land policy and in mobilizing public opinion to sustain land reform measures. Land reform is necessary, as without distributing land it is not possible to remove poverty in rural areas. It should not be judged from the point of view of productivity but it should be judged on the ground of social-justice and economic conditions of people obtaining in India.

REFERENCES

1. V.I. Lenin, *The National Liberation Movements in the East*, Foreign Publishing House, Moscow, 1962, p. 255.
2. Chester Bowles, *Making of a Just Society*, Universiiy of Delhi, Delhi, 1963.
3. P.C. Joshi, *Land Reforms in India*, Institute of Economic Growth, Allied Publishers, Delhi, 1975, p. 38.
4. United Nations, *Economic Bulletin for Asia and Far East*, Vol. VIII, No. 3, 1957, p. 73.

5. M. Raza, Land reforms and land use in Uttar Pradesh, Symposium on Land Use in Developing Countries, Aligarh, 1972, pp. 204-211.
6. A. Singh, The protection of under Raiyats in Bihar, *Yojana*, Vol. XX, 22, 1976, pp. 43-46.

17

Agricultural Regions of India and the World

SECTION ONE

INDIA

IN RECENT decades due to the rapid growth of population, the dearth of food grains has been highly felt in India. Hence it is essential to measure the productive capacity of land in different regions of the country. In this sense we should measure the amount of 'territorial differentiation of agriculture' in order to search the way for increasing agricultural productive capacity in different regions. Although the Indian Council of Agricultural Research, National Council of Applied Economic Research, Ford Foundation Group, Food and Agricultural Ministry and other research institutions of India have talked a lot on this issue but the contributions made by Mukerjee,[1] Ram Chandran[2] and Gupta[3] are praiseworthy in this direction.

BASES OF DIVIDING INDIA INTO AGRICULTURAL REGIONS

The role of climate is great in creating the regional differentiation of agricultural production. Among various climatic factors rainfall has a deterministic position. Hence before dividing India into agricultural regions it is necessary to search homogeneous climatic regions, which help us in demarcating macro or larger agricultural regions. On this bases we may divide India into the following four macro-agricultural regions :

(1) Himalayan Zone—1000 to 2500 mm rainfall,
(2) Dry Zone of the West—Less than 700 mm rainfall,
(3) Sub-Humid Zone or Central India—700 to 1250 mm rainfall; and

(4) Humid Zone of North-East and South-West India—Above 1250 mm rainfall. In order to divide these macro units into meso and micro regions or smaller units one should consider the landforms, soil and regional differentiation of climatic factors of different parts of India. In this way the meso-agricultural regions of India are as follows :

	Number of Micro-Regions
1. *Himalayan Zone* :	
1.1—Western Himalaya	2
1.2—Eastern Himalaya	2
1.3—Eastern Himalaya	2
2. *Dry Zone* :	
2.1—North-Western Dry Zone	6
2.2—Peninsular Dry Zone	5
3. *Sub-Humid Zone* :	
3.1—Upper and Middle Gangetic Plains	8
3.2—Sub-Humid Peninsular Plateau	10
3.3—Sub-Humid Coastal Region	6
4. *Humid Zone* :	
4.1—River Basins and Deltas	4
4.2—North-Eastern Plateau Regions	4
4.3—Mountains and Plateau of the East	5
4.4—Western Coastal Region	6

In the end with the help of above mentioned macro and meso-agricultural regions and crop combinations, 60 micro-agricultural are going to be presented which are the real agricultural regions of India (Fig. 17.1).

MACRO, MESO AND MICRO AGRICULTURAL REGIONS OF INDIA

Himalayan zone : It includes the States of Jammu and Kashmir, Himachal Pradesh, mountainous areas of Uttar Pradesh, Eastern Sikkim, Darjeeling district of West Bengal and the north-eastern frontier region of Arunachal Pradesh. western Himalaya is more dry in comparison with eastern Himalaya, where rain falls 1000 to 2500 mm respectively. Only 7 per cent land of the Himalayas is suitable for agriculture.

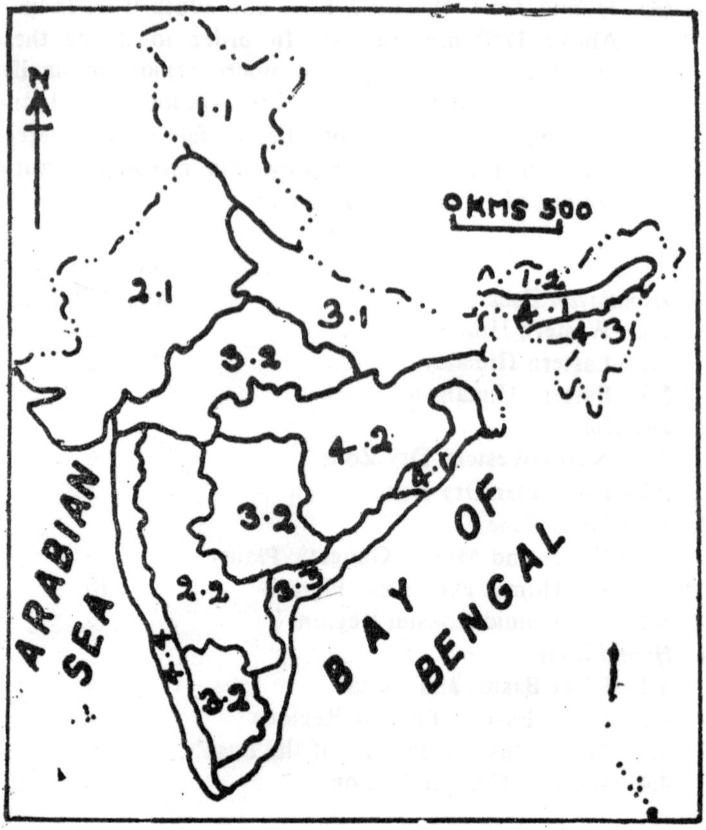

FIG. 17.1 : *Agricultural Regions in India*

Western Himalaya agricultural regions : It includes Jammu and Kashmir, Himachal Pradesh and the mountainous tracts of Uttar Pradesh. It has been divided into two micro agricultural regions which includes the whole of Kashmir except the Valley, Himachal Pradesh and Uttar Kashi, Dehradun, Tehri-Garhwal, Chamoli, Pithoragarh, Almora and Nainital. There is a dearth of agricultural land as most of the people are doing cultivation of fields besides ranching of cattle. In the extreme north only ranching is prevalent. Due to frost rain, the cultivation begins only after the melting of snow at the end of April. The important crops of Rabi are: wheat barley, gram, and

Kharif crops are: maize and rice. According to production the chief crops are rice, maize and wheat.

The valley of Kashmir : The valley of Kashmir is famous for producing rice and fruit with the help of irrigation. Water for irrigation is fetched from local minor streams. Green manure is generally used in the field. Maize is raised from peat soil of the valley. Pastoral industry is also found here and there. Crops are also grown on the water surface of lakes, especially on wooden raft. According to production the chief crops of the region are wheat, maize and rice. Besides this horticulture, sericulture, ranching of livestock and fruit gardening of apple, grapes, khumani and akhrot are also produced, which are exported to other parts of the country.

Eastern Himalayan agricultural region : It includes Darjeeling district of West Bengal, besides Sikkim, Bhutan and Assam. Here the mountain is simple structure in comparison with Western Himalaya, but it receives high amount of monsoonal rain. It can be divided into two parts.

Darjeeling-Sikkim Himalaya : Terraced cultivation is practised here. Rice, jowar, bajra and buck-wheat are raised on the upper part, whereas tea on the hill slopes.

Assam Himalaya : It is a hilly region in which Subansri, Siang, Lohit and Tirap sub-division of frontier area are located. Here shifting cultivation is prevalent where after four or five years of cultivation the fertility of the land depletes and then the search of new land has been found to be essential for raising crops. Terraced rice field is the chief characteristics of the agriculture of this region where rice is the dominant mono-crop raised everywhere.

Dry agricultural region : This is the region where people feel scarcity of water in almost all seasons for raising crop.

North-Western drier region : It includes the plains of Punjab and Rajasthan, Aravalli mountains and the Kutch Peninsula. Except the Punjab there is a least possibility of the availability of water, hence diversified cropping pattern is seen everywhere.

Desert Plains of Rajasthan : The extreme western part of Rajasthan especially in Jaisalmer, Bikaner, Jodhpur and Churu districts where the annual rainfall is less than 200 mm are included in this region. The chief crops are jowar, bajra and oil-seeds.

Semi-desert plains of Rajasthan: It includes Jhunjhunu, Sikar, Nagaur, Ajmer, Pali, Jalore and Barmer districts of Rajasthan, where annual rainfall ranges from 200 to 500 mm. The chief crops of this region are jowar, bajra, wheat and oil seeds.

Southern plains of Punjab and Haryana: It includes parts of the Punjab, Haryana, the district of Ganganagar in Rajasthan, and Mathura and Aligarh of Uttar Pradesh. Here annual rainfall varies from 200 to 600 mm and the chief crops are gram, jowar, bajra and wheat.

Northern plains of the Punjab and Haryana: It comprises the districts of Gurdaspur, Hoshiarpur, Amritsar, Jullundhur, Ludhiana and Patiala in the Punjab and the districts of Ambala and Karnal in Haryana. Here the annual rainfall ranges from 500 to 1000 mm and the chief crops are gram, wheat and millet.

Aravalli mountainous region: It comprises the districts of Alwar, Bharatpur, Jaipur, Sawai-Madhopur, Tonk, Bundi, Bhilwara, Udaipur and Dungarpur in Rajasthan. Here the average annual rainfall is 800 mm and the chief crops are maize, wheat and gram.

Saurashtra and Kutch Peninsula: It includes part of Gujarat State except Surat, Bharauch, Baroda, Moghra and Kaira districts. The average annual precipitation is about 700 mm and it is a region of black lava soil where groundunt, millet and cotton are raised.

Dry Peninsular region: It includes the plateau region of Maharashtra, Karnataka, Andhra Pradesh and Coimbatore. It lies in rain shadow zone of Western Ghat mountains and hence it is a dry zone. The chief soils are red and regur, which are in most cases unproductive. The chief crops of this region are wheat, millet, oil-seeds and cotton.

Narmada-Tapti doab : It includes the western Nimar district of Madhya Pradesh and Dhulia, Jalgaon and Nasik districts of Maharashtra. The crops associated with this region are millet, cotton and groundnut.

Upper Godavari basin : It comprises Aurangabad and Ahmed-

nagar districts of Maharashtra. The significant crops of this region are millet, wheat and oil-seeds.

Upper-Krishna and Bhima Basin: It comprises the districts of Poona, Satara, Sholapur and Sangli of Maharashtra. The main crops are millet, groundnut and oil-seeds.

Tungbhadra Basin: It includes the Kurnool district of Andhra Pradesh and Raichur, Bellari, Dharwar, Shimoga and Chitradurga of Karnataka. Tank is the main source of irrigation and the crops grown are millet, groundnut and cotton.

The Plateau of Rayalaseema: It is a zone of laterite soil which comprises the districts of Anantpur, Cuddapah and Chittoor of Andhra Pradesh. Rice is grown on foothill zones and the other crops are millet, rice and groundnut.

Sub-humid agricultural region : It has been divided into three meso-agricultural regions:

Upper and the middle Gangetic plains: In upper Gangetic plains, the doab of the Ganga-Yamuna and the plains of Rohilkhand are highly irrigated region. Half of the net sown area has been found to be irrigated and the diversity in crops are available. Here wheat is sown with special care besides sugarcane, rice, millet and gram.

We can divide the Upper Gangetic Plains into four micro-agricultural regions which are as follows:

Upper Ganga-Yamuna doab: It includes Saharanpur, Muzaffarnagar, Meerut and Bulandshahr districts of Uttar Pradesh, where wheat, sugarcane and rice are the main crops.

Lower Ganga-Yamuna doab : It includes Agra, Itawa, Farukhabad, Kanpur, Fatehpur and Banda districts of Uttar Pradesh. The crops of the region are wheat, millet and gram.

The plains of Rohilkhand: It comprises the districts of Bijnor, Moradabad, Badaun, Rampur, Bareilly, Pilibhit, Shahjahanpur, and Hardoi of Uttar Pradesh. About one-fourth cultivated area is irrigated and the significant crops are wheat, rice and gram.

Northern terai region : It includes the districts of Sitapur and Lakhimpur which lie in the Himalayan foothills of Uttar Pradesh. The soils of this region have greater moisture holding capacity in spite of the lack of adequate rainfall. Rice, wheat and maize are the chief crops of this area.

Middle Gangetic Plains include eastern Uttar Pradesh and the northern plains of Bihar. It is a highly productive region of India where soils and climatic conditions are favourable for agricultural production throughout the year. This region is situated between the moist Ganga delta of West Bengal in the east and drier Gangetic Plains in the west. The average annual rainfall is 1100 mm. The important crops are rice, wheat, maize, sugarcane and gram.

Eastern U.P. and Bihar region: It comprises the districts of Gorakhpur, Azamgarh and Gazipur of U.P. and Champaran in Bihar. Rice, wheat and sugarcane are the chief crops of this region.

Tarai region of U.P. and Bihar: It includes Gonda, Bahraich and Basti districts of U.P., and Saran, Muzaffarpur, Darbhanga and Saharsa districts of Bihar. Rice, wheat and maize are the crops which grow here.

South Gangetic plains: It comprises the districts of Mirzapur, Allahabad and Varanasi of Uttar Pradesh and Shahabad, Patna, Gaya, Monghyr and Bhagalpur of Bihar. Rice, gram wheat are the chief crops of this region.

The Plains of Awadh : It includes Gorakhpur, Deoria, Bhatni and other districts of central Uttar Pradesh. The crops are sugarcane, rice and gram. The soil is fertile and partly irrigated.

Sub-humid peninsular plateau : The significant crops of this region are millet, wheat, gram, cotton, rice, oil-seeds, groundnut and sugarcane. There are 10 micro regions of crop association i.e; the plateau of Malwa, the Valley of Tapti, South-Eastern Plateau of Maharashtra, Northern Telengana, Southern Telengana and Karnataka Plateau, Northern Bundelkhand, Southern Bundelkhand, Bardha Basin, Tamil Nadu Plateau and Southern Malnad region. Among all these micro regions at least five regions specializes in rice, cotton, wheat, millet and gram. This is the region of regur soil which have a definite bearing on the crops of the region. In areas of red soil rice and groundnut dominates the scene instead of cotton. In Southern Bundelkhand and central Indian Plateau wheat, millet and gram grow in association. Gram and wheat are the crops of Rabi season whereas millet and gram are the crops of Kharif season. In Southern Telengana the combination of millet emerges with rice and oilseeds.

Sub-humid coastal region : Rice is the main crop of this meso agricultural region. It extended from Srikakulam in coastal areas of Andhra Pradesh to Tiruneveli district of Tamil Nadu. Some difference in cropping pattern is observed in Guntur and Nellore districts where millet is more important than rice. Alluvial soil is found in the coastal region which has the facility of canal irrigation. The important crops are rice, millet, oilseeds and pulses. It has six micro-agricultural regions, i.e, upper part of east coast, Krishna and Godavari delta including parts of Andhra coast, northern coast of Tamil Nadu, southern Tamil Nadu and northern Gujarat coast.

Humid agricultural region : In humid region the diversification of crops are not so important as in drier areas. Due to higher amount of precipitation rice is a permanent crop besides wheat and millet. On south-western Malabar coast spices, coriander, tea, rubber and coconut dominate the scene, but in north-eastern part, tea and jute are important crops of the region. It has four meso regions i.e:

Deltas and river valleys : It stretches from the valley of Brahamaputra to Mahanadi delta via the Ganga delta. It has thick layer of alluvial soil where rice and jute are the chief crops of lowlying areas and tea on mountainous slopes. It has six micro regions such as, upper Brahmaputra valley, lower Brahmaputra velley, Jalpaiguri district, Malda district and Bhagirathi delta of West Bengal, Orissa coast and the southern part of the Ganga delta.

North-eastern peninsular plateau region : In this meso-agricultural region upper Narmada valley, mountains of Mahadev and Maikal plateaus of Baghelkhand and Chota Nagpur, upper Mahanadi Basin, plateaus and mountains of north Orissa, mountains of Bastar and Koraput, mountains of South Orissa and Banganga Basin are included. 75 per cent land of total net sown area is under rice.

Eastern mountains and plateaus : Here shifting cultivation is prevalent after burning of forest and due to torrential rain the problem of soil erosion is highly felt. Rice is the mono crop in the plateau region of Manipur and mountainous areas of

Mizo hills. Rice is again a mono-crop in Shillong Plateau, and more or less the same cropping pattern is available in Mikir, Cachar, Khasi and Jaintia hills of this region, on hill slopes tea gardens are found but on the waterlogging plains rice and jute are chiefly grown.

Western coastal plains : Linear coastal plains of Western ghats is the chief characteristics of this region which may be divided into the higher slopes of Western ghats, plains of alluvial soil in the middle and sandbars on the coast. High temperature, higher rainfall and evergreen forest are the characteristics of this region. Plantation agriculture is found in the southern part. The chief crops are tea, coffee, spices, rubber, coconut and cashewnut besides rice. It has six micro-agricultural regions such as coastal regions of Khambhat, northern Konkan coast, southern Konkan coast, Karnataka coastal region, northern Malabar coast and southern Malabar coast.

SECTION TWO

WORLD

The division of agricultural regions should be made on the basis of association of crops and livestock, characteristics of agricultural practices, availability of labour, capital and social organization, mode of consumption of agricultural produce, use of agricultural implements and human settlements, etc. According to Whittlesey there are thirteen agricultural regions of the world which are as follows: (Fig. 17.2):

(a) Nomadic Herding Region
(b) Livestock Ranching Region
(c) Shifting Cultivation Region
(d) Rudimentary Sedentary Region
(e) Intensive Subsistence Tillage with Rice Dominant Region
(f) Intensive Subsistence Tillage without Rice Region
(g) Commercial Plantation Tillage Region
(h) Mediterranean Agricultural Region
(i) Commercial Grain Farming Region
(j) Commercial Livestock and Crop Farming Region
(k) Subsistence Crop and Stock Farming Region
(l) Commercial Dairy Farming Region
(m) Specialized Horticulture

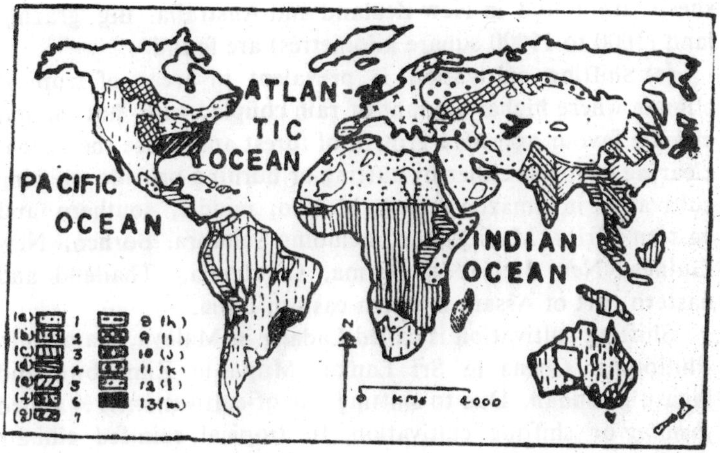

FIG. 17.2: *Agricultural regions of the world*

(a) Nomadic herding is prevalent in the drier areas of Africa and in the Tundra grass lands of Asia. The local tribes Khirgiz and Lap people are engaged in this occupation. These people derive their livelihood from livestock in the form of milk, meat, leather, ranching, tent making, along with the arms made of bone. In this way livestock acts as the pivotal point in shaping the economic system of the region. Sheep, goat, camel, cow, ox, yak and reindeer are predominantly found. Thorny bushes and small grass are suitable for rearing goat, but in drier areas sheep and camel are more prevalent. On the highlands of Tibet-yak, and in Tundra region the abode of reindeer and caribou are found.

(b) Livestock ranching is found in the desert and semi-desert environment where European people have settled. Livestock ranching dominates the scene in western part of North America, Australia, New Zealand, and parts of Patagonia desert, Chako region, the plateau of Brazil in South America and the south-eastern plateau region of Africa. In north-eastern part of Caspian sea in U.S.S.R. livestock ranching has begun, where nomadic herding was found in earlier period.

In this region livestock rearing is important among which American region specializes in cow and ox whereas goat and sheep are found in New Zealand and Australia. Big grazing land (1000 to 15000 square kilometres) are found.

(c) Shifting cultivation is prevalent in areas of tropical climate where higher amount of rain coupled with hot humid climate favour excessive growth of forest and hence the casual clearing of forest either by cutting or burning provide shifting cultivation in Amazon region, Mexico, middle, southern and eastern Africa, Madagascar, including Sumatra, Borneo, New Guinea, New Hebrides, Burma, Cambodia, Thailand and eastern part of Assam in north-eastern India.

Shifting cultivation is called Ladang in Malaya, Caingin in Philippine, Chena in Sri Lanka, Milpa in Zimbabwe and Ngasu in Sudan. Due to shifting site of cultivation it is called *jhuming* or shifting cultivation. In tropical rain-fed climate the leaching of soil, quick exhaustion of the fertility of vineer layer of soil and quick growth of vegetation favour shifting cultivation. The chief crops of this region are maize, jowar, bajra and paddy. Maize profitably is produced in Amazon and Mexico, jowar and bajra in Africa and rice in south-eastern Asia. Besides these side crops such as groundnut, beans, tomato, cucumber, plantain, maniyak, yam, kasaba and sugarcane also find their place. The selection of site for cultivation is the greatest problem in this region.

(d) Rudimentary sedentary agricultural region is found in areas of hot humid climate, but the shifting site of cultivation is not found. Here due to relatively high density of population permanent form of rudimentary agricultural systems have developed. The chief crops are spices and rubber. In areas of lesser rainfall and dry climate this agricultural system in commonly practised, especially in areas of alluvial soil or valley .plain. The livestock of this region are horse, sheep, goat, cow and oxen.

(e) Intensive subsistence tillage with rice dominant region is found in India, China, Japan, Burma Cambodia, Vietnam, Korea and the whole of East Indies. It is known as Sawah agriculture. In these areas high density of population (400 to 2000) per square kilometre, smaller fields, high monsoonal rain and annual inundation of field by the overflow of rainfed streams favour intensive subsistence farming. The chief crops are rice, maize, fruits, beans, sugarcane and jute. Annual rainfall varies from 150 to 1200 mm. Four consecutive seasons are summer, rainy, winter and autumn. Settlement is compact and semi-sprinkled in most cases and people rear cows, buffalo, oxen, goats, pigs and fowl.

(f) Intensive subsistence tillage without rice region include whole mid-western part of India, Manchurian north of Peking, parts of Laos and Cambodia. In these areas wheat, cotton, tea, rubber and cinchona are grown widely. It is known as dry-field agriculture.

(g) Commercial plantation tillage region is found in Latin America, Africa and South-Eastern Asia. Plantation agriculture is found only in smaller part of the region. The important crops are plantain, cocoa, coffee, sisal, rubber, sugarcane, coconut, and spices, etc. Plantain is chiefly grown in West Indies, coastal areas of Ecuador and mid-America. Sugarcane is a dominant crop in Cuba and the coastal tracts of Peru and Venezuela. Coffee is grown on the Plateau of Colombia and Brazil. In Africa, coffee and cocoa are grown on mid-western and eastern parts, whereas sugarcane and sisal grow on eastern plateau region and Madagascar. Coffee and coconut grow on southern plateau region of India, whereas tea is chiefly grown on Nilgiri and mountainous areas of Assam and the plateau region of Sri Lanka. Rubber is an important crop in Malaysia. On such plantation farms offices of the company, factories for preparing goods and labourers house are generally located basides plantation of crops. Cheap labour is abundantly needed especially in the plantation and harvesting of plantain, tea, rubber and coffee.

In the region of commercial plantation the problem of soil erosion, depletion of fertility of the soil, high temperature and the lack of sufficient rainfall are the main hurdles in the development of the region.

(h) In the Mediterranean region both intensive subsistence and commercial farming are practised. Here altogether a different combination of crops and livestock rearing have been found. The importance of food crops, fruits, vegetables and in livestock sheep, goat, cow and oxen are seen for the sake of milk, meat and wool. The annual rainfall varies from 45 cm to 100 cm. The chief crops are barley, wheat, olive, lemon, grapes, apple, potato, tomato, onion, beans, oats, etc. which grow in summer and winter seasons both with the help of irrigation and without irrigation.

In the Mediterranean agricultural region subsistence agriculture is found in Middle Spain, Southern Italy, Turkey and Syria whereas commercial agriculture is prevalent in California, Australia, Africa and South America. The main characteristics of Mediterranean agriculture are the multiple cropping, variations in the characteristics of agriculture and the size of farms.

(i) Commercial grain farming region is found in the grass-lands of mid-latitudinal belt. This type of agriculture is just limited to the Prairie region of North America, steppes in U.S.S.R; Pampas in Argentina and the basin of Murray and Darling in south-eastern Australia. The characteristics of this agricultural region are the specialization in one crop, excessive use of machine, large size of agricultural farms and extensive cultivation. Wheat, corn, barley and millet are chiefly grown. 60 per cent to 80 per cent land is engaged in wheat cultivation. According to the needs of family cow, hogs and fowl are also reared in this region. For the ploughing of field and sowing of seeds tractor, developed plough, disk, harrow drill, and to harvest the crop reaper, combine harvester, etc., are extensively used in this region. The size of farms vary from 500 to 2000 hectares. After harvest wheat is generally sealed in gunny bags and it goes to elevator with the help of trucks and then to the port via rail line for exports.

(j) The commercial livestock and crop farming region is found in the Ohio, Indiana, Illinois, Iowa, Nebraska, Virginia, Tennessee, Georgia, Oklahoma States of U.S.A. and over the whole of Eurasia. The maximum width of this region is 1200 kilometres which stretches from Ukraine to Finland.

(k) The chief characteristics of this agricultural region are the rearing of livestock and production of food in farm. Wheat and corn are important crops and besides these potato, cauliflower, beans, peas, turnip, coriander and sugarbeet are also grown in this region.

There is also a provision of crop rotation in between corn, wheat, pasture and hay. Hay includes alfalfa, clover and timothi. Although agricultural implements are extensively used but altogether it is a region of intensive cultivation.

Here the average size of farms are 50 to 200 hectares in which the house of the peasant, livestock, storehouse of grains and implements are located. The livestock of this region are hogs, cow, sheep and fowl.

(l), (m) Commercial dairy farming region includes north-western Europe from Atlantic coast to Moscow, Prairie region to Atlantic coast in North America including south-eastern Australia, Tasmania and New Zeland. The chief aim of such an agriculture is to produce fodder and grass for milk production including breeding of cattle wealth. Among various races of cows such as Guernsey, brown, Swiss and Ayrshire, Holstein variety gives more milk and cheese. Maize and barley are the chief crops which serve the purpose of rearing livestock. In U.S.A. the size of dairy farms varies from 25 to 100 hectares but in Denmark 50 per cent farms are below 10 hectares of size.

Ultimately it can be concluded that physical factors, technical knowledge of inhabitants of the area and the demand set by the people for certain utilization of land determine the agricultural region in any part of the globe.

REFERENCES

1. A.B. Mukerjee, Agricultural regions and geographic planning for Indian agriculture, *National Geographical Journal*, Vol. 8, 1962, p. 407.
2. R. Ram Chandran, Crop regions of India, *Indian Geographical Journal*, Vol. 38, 1963.
3. P.S. Gupta, Economic regionalization of India: Problems and approaches, Monograph No. 8, *Census of India*, 1961, pp. 101-110.

18

Land Use Planning

LAND USE planning in rural areas has received scant attention of geographers in comparison with the town planning. This can hardly be justified in a country like India which is predominantly agricultural. The planning of agricultural sector in rural areas has been carried out in a laissez-faire manner. Very few deliberate and cautious efforts have been made to improve the countryside. Yet an acquaintance of the agricultural land use would show that all development work has more or less been guided by the land resource available for cultivation. Land provided sites for various uses such as houses, pastures, water-reservoirs and agriculture. The selection was mainly guided by the physical landscape as well as the needs of the rural communities. As time passed by, these early sites were adjusted to meet the new demands.

According to the Survey and Settlement Report, the Permanent Settlement Act of the 19th Century in India, vested the local landlords in the territories of Bengal Presidency with tremendous power. Afterwards, the initiative for all development works shifted to the landlords of the village. Some landlords were negligent, while others discharged their responsibilities with full courage and resources available, but in all cases they were imbued with more self-interest than public good. This situation continued until the acquisition of Zamindari Act of Bihar in 1951.

Thus the planning percept of land use has legal significance. On a regional or local level in exercising planning and zoning process, it has reasonable consideration to the character of each plot of land and its suitability for particular uses. The land-use survey should be planned and programmed in advance. The purpose of the survey should be identified and the amount of information and the degree of detail should be balanced against their ultimate use. At present this task has been given to the Community Development Programme in India.

With the introduction of community development programme all powers and functions of rural planning have been laid on the shoulder of Block Development Officer. He is the chief executive officer of the Five Year Plan Projects. He has the necessary fund and staff to assist in the development work. In some areas the community development programme has been in operation for well over 15 years, yet the overall development in agricultural landuse is far from satisfactory. It is suggested that rural planning especially in the field of agriculture can be better executed with the help of land use survey of the area.

PLANNING PROCESS

The development of land use planning is not recent in its origin but its practice is truly recent. A land use map gives a through and clear picture of land to the planner for the determination of future use and planning the agriculture sector to maintain the land potentials. In this way, land use planning is concerned with the future use of land and the changing demands of the society. In words of Prof. Stamp, 'land planning is in essence the determination of the optimum use of every acre of land which must be elastic and can change from time to time to adopt the changing conditions'.[1] "Land use planning is more than a compound of architecture, engineering, public administration and social sciences. It is a new and engineering discipline with its own scientific and objective approach to its central theme, the use and development of land".[2] It suggests the future potentialities of land also.

For planning purpose land use data can be amply used to determine crop specialization, crop combination, commercial markets, to locate settlement and institution such as church and school, or for zoning purposes. In zoning studies, with which we are concerned here, it is essential to know the amount of land used for various purposes. Zoning is based on the facts of actual use of land and that has greater validity than that based on opinions only.

The land use map represents the conditions on a given date, and its validity becomes progressively less with each change in use of property. Thus, it is important to keep this graphic and statistical information current. Periodic revision is necessary.

Land use planning should start from the present position of the land and should be based on a careful objective, detailed survey and analysis of the present land use pattern whose future use must satisfy the basic needs of mankind. "There are at least six such basic needs of man which involve the use of land for their satisfaction. We may list them as food, shelter, work, recreation, movement and security. Where population is dense these become competing demands".[3] Thus land use planning includes both economic and social importance. The total supply of land resources is fixed in both quantity and geographic location. Under property rights, we have given private and public ownership of land with considerable freedom to decide how land resource should be used or misused. Demand for public action to direct land use practices seldom arises as whether there is a sufficient supply of land resources at the right locations to fulfill the needs of prospective users.

All planning, public or private, involves a process. It embraces a series of steps which can be classified in different ways but normally involve :

(a) Objectives of Planning;
(b) Formulation of a Plan; and
(c) Execution of the Plan.

(a) The goal of planning depends upon the planner's individual experience or intuition. It may be an outcome of conscious thinking of an individual, a group of persons or an agency. The planning may be good in one sense and bad in another. It may be fixed for all time or be accepted as temporary objectives subject to future revision. Thus, the goal of planning always plays a vital role in giving proper direction to the planning process.

(b) The formulation of a plan varies with individual cases. Mostly land use planning involves four steps :

(i) researches and studies concerning the nature of the problem;
(ii) analysis of the alternative solutions that can be used;
(iii) choice of combining other essentials of the plan, and
(iv) the formal drafting of the plan in written or graphic form.

Planning efforts frequently go astray as they are based more upon wishful thinking than upon a sound understanding of the facts. But the working knowledge of the planning shows that it is a prerequisite for successful planning. Area planners may gather a wide range of data and analyse them according to their own choice. But in rational decisions of planning, it may be impossible to acquire all such information at a time. In such a situation, the planners should always base their decisions on findings that represent a reasonably accurate picture of the present situation in hand.

(c) Planners are always interested in effectuation or carrying out of their plans. It is at this stage that a plan can prove itself successful or a visionary master blue-print of utopia. Good administration, successful training and political stability are pre-requisites for plan execution. Besides this, cordial relationship between the government and the people is essential because continued political and financial support by the state is necessary for the execution of land use planning proposals.

ROLE OF LAND USE STUDY IN PLANNING

On the basis of detailed land use survey, land use planning need not be emphasized in India as the pressure of population is continuously increasing on land. Several schemes have been launched to increase the agricultural production in various parts of the country, to introduce new crop combination and ration of crops, to substitute better variety of crops and to improve agricultural techniques. The land use problems necessitates revision in terms of long range planning. Land use planning requires the optimum as well as multiple use of the land. The land should be used so as to satisfy the needs of the nation to the best possible extent. This is not against the exercise of freedom in choice of work, nor is it for the development of private enterprise; but it should not act as such as to come in direct conflict with the national interest.

In this way, land use planning is needed to determine the optimum use of every piece of land in the country. This may change from time to time, place to place or in war time. The land use planning must, therefore, be a dynamic force rather than a static one, it must be in accordance with the dynamic needs of the society. Any change in the character of land use by land reform measures should not be judged on the basis of economic advantage of some handful persons of the community but it must encompass the national interest.

With the help of land use study we can also indicate the vacant land, land capable for reclamation, fragmentation of holding, marshy and unirrigated land and after straightening the streets and lanes in the market and village we can get extra-bari land for the production of food to the increasing population. In this way the relocation of settlements is necessary on unproductive vacant land and that also ensures the planned development of settlements as predetermined site. This will help in saving the productive bari land. Fragmentation of land-holdings can also be checked by consolidation of smaller plots. It also indicates the location of usher land and the marshy ground. After providing irrigation and adequate use of fertilizer the usher land can be reclaimed for better production of food crops. After filling the marshy land with the dirts of the village and roads, such lands can also be partly reclaimed, and if any region lies in the flood affected zone of a river, then after construction of suitable dam such land also can be reclaimed.

Land use study should not be made for a statics time period but the study of its temporal changes is necessary for future use. Thus the study of land utilization is of immense value for planning the agricultural growth and regional development.

In Bihar state per capita net sown area is only 0.35 acre which is perhaps quite insufficient to fulfil the required needs of every inhabitant fully. The increase in population needs homestead land, roads, canals, and other amenities and it means more demand for present use of the land. Thus the reduction of per capita net sown area and cultivable land will naturally occur. The above condition shows that the scope for the extension of cultivation is limited, but the reclamation of forest land, barren and uncultivable waste fallows, permanent pastures, orchards and cultivable waste lands may help us in the extension of agricultural economy. All these give idea to the planner that with the help of better manure, seeds, irrigation and improved means of cultivation one may bring these lands into our agricultural use for feeding the teeming millions. In real world situation, the intensive cultivation of land is found only near the settlements while on the outer margins of the village, the cultivation of land is extensive in nature. Thus, if we provide condition for dispersal and relocation of settlements and the inhabitants to their respective distant fields with proper security arrangements, the provision of such an ideal condition must follow the intensive cultivation raising number of crops in one year from the present use of single cropped land. In this way we can say that there is a vast scope for extending our cultivation through land use study and planning but the only prerequisite is that of our incentive to do so.

GENERAL PROBLEMS AND PROSPECTS OF LAND USE PLANNING

To maintain the land capability and high nutritional level of inhabitants land us planning is necessary. Our agricultural production is low in comparison with the increasing population, thus how to boost up agricultural production in different categories of land is the chief problem in the entire region, so that balanced regional economic development may also be established.

For example, in Bihar the northern part is fertile plain but the havoc of flood each and every year, irregular monsoon rainfall, lack of adequate irrigation facilities and high population density make the residents poor in most cases.

In recent years, the attention of the government to assured irrigation facilities and self-reliance with the help of private irrigation management and the use of need based chemical and compost fertilizers, have raised the general productivity of land.

On the other hand, in Chota Nagpur Plateau, rugged massive forest clad terrain restricts the high density of population and stony wastes are unsuitable for raising crops in the region.

While Chota Nagpur plateau is rich in mineral deposits which raised the standard of living. The inhabitants have now greater intention to practice scientific cultivation of fields and produce high yielding varieties of rice, wheat and fruits in the southern half of Bihar.

Thus, the above discussion reveals that the State of Bihar is in a position to achieve the self-reliance in food if the government is actually helpful to the people on the front of food production.

FARM PLANNING FOR BETTER LAND USE

To provide food, cloth and shelter to each and everyone it is necessary to plan our farm land scientifically. Indians live in villages and 80 per cent people are rural, thus the farm planning on cooperative basis is necessary to ensure food, cloth and shelter to everyone.

Management: The cooperative farming should be managed by the agricultural experts of the government who should be a trained fellow either in agricultural science or an arts graduate of some *kisan* family. He should be a responsible person and his service should not be continued after three months of time unless the fellow gets the good faith of the cooperative farmers on the basis of secret ballot.

Labourers should be engaged on the basis of daily payment according to skill and capability in a particular service.

Chakbandi: For farm planning it is better to make *chakbandi* of the dispersed and smaller plots of field of the entire cultivable land of the village as the basis of 50 acres of each farm for 50 workers of the village. *Chakbandi* should be made after evaluation of the land according to its price and productivity.

Before going into *chakbandi* it is necessary to finish the contouring of the entire village on the basis of plane table survey and dumpy level. This gives us the slope category and drainage profile of the village land in cordinal directions. If the size of the village according to its cultivable land is bigger than the planner may make the farms of different sizes and cost involved in Rupees in each category are as follows:

TABLE 18.1 : **Farms of different sizes and cost involved**

Farm size in acres	Cost involved in rupees
50	50,000
100	100,000
200	200,000
300	300,000
1000	1,000,000
2000	2,000,000

Electricity : Government should produce more electricity in different zones of the state with sub-power stations and this will require the installations of smaller generating sets in the courses of different rivers with the creation of smaller dams through subversion canals which have trivial purposes of electricity generation, irrigation and fish rearing.

Fencing : Fencing should be done with shisham and semal trees so that semal may help in our match factory and shisham will solve the problem of furnitures such as chair, table, cot, plough, etc.

Soil : The examination of land capability according to soil categories is necessary. The examination of soil PH and texture and its proper diagnosis for the utilization of particular fertilizer in a particular farm as the sampling basis is essential before going to sow a certain crop. This necessitates the establishment of soil testing stations in each and every Panchayat.

Cropping pattern : After examining the soil quality the cropping pattern should be determined and each and every farm should produce at least three crops in a year in rotation. The farmers should produce fruits, vegetables, foodgrains and especially sugarbeet, peas, beans, varieties of pulses, oil-seeds and summer paddy and maize etc.

Organic fertilizer : The peasant should construct three pits on the side of livestock farm, which are to be used, as follows: The one should be used when other has ceased to be used and the third should be under formation for three months. To solve the shortage of fertilizer peasants may supplement the organic fertilizer with chemical fertilizers.

Dusting, liquid spraying and top-dressing are necessary in case of new leaves for better produce and for saving the crops from rusting and fatal pests and diseases. It also helps us in raising the production of different crops.

Garden : Fifteen acres of farm should be reserved for mangoes and seasoned fruits like citrus, plantain and papaya, etc. (50 plants each in a 50 acres farm). With the help of growing varieties of seasoned fruits we may develop juice factory, jam factory, etc.

Sugarcane cultivation : 5 per cent of land in a farm should be reserved for sugarcane cultivation and with the help of this we may establish smaller sugar factory in a group of every ten villages.

Cotton cultivation : In every 1000 acres of farm we should be reserved the 15 acres of higher land for cotton production and after three pluckings, we may destroy the cotton plant, sometimes in summer season when all the plants have gone dry in the scorching heat of summer.

Agricultural markets : To reach the products and manufactured goods of rural areas into the market it is necessary to have a lateral road connecting every village to the nearby town where the government purchasing godown or agricultural market is situated. On the other hand, government should immediately provide fertilizer, cement, asbestos sheet and agricultural implements and pesticides barter basis to the peasants on balance money.

Crop bank : Government should construct on the zonal basis crop bank containing the share of each and every farm and being controlled by the government without jeopardising the interest of the farmers. It is optional for the farmers to keep their grains into the grain bank, and the godown should be scientifically well maintained. No transactions of money should be made here and peasants should obtain bank draft.

Industries : Agricultural goods based industry should be established in the central settlement of cooperative farmers to process milk for dry powder and varieties of other things like butter, ghee, etc. from fruits jelleys, chatni, drinks, etc. and from grains wheat, corn and potato chips and varieties of other preparations can be made. Husking machine should also be established on zonal basis and the excrete of husking material may be utilized as fodder, which should be purchased by the peasants of the village.

Vocational school : Vocational school should be set up in the centre of six villages where after passing the middle (seventh class) the training of carpentry, smithy, bottlemaking, spinning of cotton thread for cloth weaving, different agricultural operations, boat making, shoe making, tanning, poultry keeping, music, dancing and several other odd trades should be given to the student before going into the service of the society.

Higher education : The age should be fixed for obtaining higher education and only after getting vocational training one should be admitted for specialized training in higher education.

Employment : To employ the educated youth there should be radio-manufacturing centre in a group of every six villages, mini-steel plant should be established to fulfil the smaller demands of rural inhabitants. This will ensure the employment to more people with higher wages. As regards teachers from primary school level to the post graduate one and the government servent, there should be an arrangement for housing, and the transfer of all employees should be made necessary after completion of three years of time in the same cadre or promoted post.

Tackling the problem of flood : Overspill of water from the channels due to heavy monsoonal rain plus excessive melting of snow on the Himalayas are the causes of flood in North Bihar. During flood, most of the cultivated lands get submerged, crops and settlements are washed away and thousands of poor people are rendered homeless. Certain zones have greater flood problem than others in the State of Bihar.

To remove the problem of flood, government should make a diagnostic survey of the entire area. For the construction of dams according to overflow capacity on firm site, selection at required contour intervals in the channel of several streams, is necessary. Such streams should have numerous diversion canals of 8 feet width required for the regulation of water. Arrangements should be made to store the water in different artificial reservoirs keeping in view the need for irrigation felt by the local farmers. With the help of these dams the government may establish hydraulic power stations and in the reservoirs the fishing industry may be developed. This will also develop the fish tinning, Zeora rearing and Makhana raising business in different reservoirs at certain stages of the channels.

On the whole, farm planning depends upon the sincerity of the cooperative farmers which is lacking in common Indians. If it will be adhered to, it will solve the problems of unemployment by raising job opportunity, fulfil the shortage of food by better production, solve the problem of cloth by cotton produce and shipping and weaving the same in leisure hours. The plantation of shisham helps in constructing houses and making furniture ; while more produce helps in reaching the desired goal of the community in every walk of life.

Defects in the present land use : It is quite evident from the analysis of the pattern of land utilization in rural areas of Bihar that with the present methods of land use, the potentialities of land resources are not being properly exploited, rather there is a great misuse of the land resources. This necessitates proper planning, keeping in view the defect of the present land use and the principles of the maximum and optimum unutilization.

Planning for the future : The foregoing discussion of the land utilization in rural areas of Bihar indicates that even the cultivated land is not being exploited to its maximum and needs the following improvement :

(a) *Increase in the double-cropping* : The double-cropped area which is only 25 per cent at present be extended to whole of good quality and medium quality land, with the help of better irrigational and manuring facilities.

(b) *Increase in yield* : Per hectare yield is far from satisfactory and be increased considerably by maintenance of soil fertility, scientific crop rotation, increase in irrigational facilities, improvement in farming operations and use of high strains of plants.

Change in the cropping pattern : The problem of better use of the land is not only of increasing the agricultural production, but also of having a sound cropping system, which may meet the multifarious demands of the growing population. In order to meet the deficiency of calcium, fat, vitamin A and C in the diet of the people, leafy vegetables such as spinach, amaranth, coriander, drumstick, mint, radish, etc. be grown on the land of good quality. Milk is the main source of various deficient nutrients, the production of which will immensely increase, in case the barren land is reclaimed for pasture and grassland.

The villagers should leave their traditional land use practice in respect of changing man-land ratio and scientific development. The government and peasants should provide irrigation to each and every field. They should practice co-operative farming for the integrated area development and better training facilities in bigger cooperative farms having good transport network and communication lines in rural areas for the movement of machines and men. In some areas these movements have already been followed by the peasants under crash programme and grow more food campaign. Thus under such a changing land use practice, it is quite necessary to note down the traditional pattern of land use and remove the old system of cultivation for feeding the millions of growing people of our villages and towns so as to raise their economic status and help them to achieve self-sufficiency.

Till now land use planning is mostly conducted on paper, and nowhere it has been realized in the field by the administrators. Thus, the planning body should take suggestion from the land use experts for its diagnostic survey. The application of land use in the field needs reconnaissance survey of the region, its resources assessment and management on scientific lines which has really helped the peasants in land use planning.

The farmers in Bihar are poor. They cannot spend huge money on agricultural instruments and fertilizers. The population density is high which needs more job opportunities, but without helping and solving the problems of these teeming millions, no land-use-planning can be successful in the State of Bihar. The need based financial assistance alone can solve these problems en masse.

REFERENCES

1. L.D. Stamp, *The Land of Britain : Its Use and Misue*, op. cit., p. 426.
2. J.N. Jackson, *Surveys for Town and Country Planning*, Hutchinson University Library, London, 1963, p. 1.
3. L.D. Stamp, *Applied Geography*, Hammondesorth, 1960, p. 64.

19

Conclusion

LAND use as a comparatively new branch of economic geography is intimately associated with demand for a particular use of the land. Land use changes to meet the variable demands of land by the society inspired by technological changes. In this way land utilization involves an examination of the natural factors affecting both the harnessed and the potential productivity of the land. Land economics and land utilization also have a close relationship. The efficient use of land depends on the capacity of man to utilize the land and the systems of farming, land tenure, size of agricultural holdings and institutional framework prevailing in the region.

Land use study involves principles, techniques, controlling factors, classification, capability measurement, resource appraisal conservation, land reforms and planning of land use. L.D. Stamp of Britain was the pioneer in this field. Since then some of the developed nations like the U.S.A., U.S.S.R., the U.K., Denmark, Italy, Cyprus and Poland have completed extensive soil surveys and mapped out land use, while in developing and under developed countries the same is wanting as yet. Hence, in a way to develop our agricultural economy the study of land use is pre-requisite rather than a necessity. Indian geographers have made some headway towards land use study but most of the studies pertain to smaller regions which is of no importance in comparison with the vast territorial expansion of India.

It has been found that the land use is controlled by two sets of variables mostly physical and cultural. Physical factors include topography, climate, soil and vegetation which provide the regional framework and limits the use capabilities. Human factor includes the length of occupance of the area, demographic and socio-economic conditions, institutional framework and the technological levels of the people which determine the extent of land utilization. Locational factors also have their influence on land use of a particular area by way of transportation lines and bringing capital and labour to the land. This involves location problem in the study of land utilization. Similarly institutional factor includes governmental, industrial, educational, family and subordinate secondary institutions such as constitution, legislation, political parties, civil service systems, tax regulations, tariff control and social security which are the determinants of land use in any region.

So far as the principles of land use is concerned, the consideration of the nature of the land is a vital issue. For this fact, one use of the land excludes other, such as agricultural use excludes urban, recreational and transportation use.

The area of earth surface is fixed and in no way it can be extended. This shows that the superiority of demand should be the first principle in any scheme of the utilization of land. Secondly the character of land such as plains, plateaus and mountains cannot be replaced except by diastrophism. Hence, one should satisfy oneself whatever may be the character of land. Similarly, land use is guided by ecological control of variables, productive potential and human values of the land. Thus, the locational advantage and the demand for a particular use on certain land is a more important guiding principle of land use than others.

In delineating the approaches and concepts of land use, it has been found that the Malthusian approach of population checks is not applicable to the present day situation. But the Malthusian concept of 'struggle for land' shows the importance of land as a factor of production in relation to the evolution of human settlements and the degrees of land utilization. Another approach of land use is given by Marx. The supporters of Marx's view, favour the 'law of diminishing returns'. They suggest that if an additional investment of capital and labour on land produced not a diminishing but an equal quantity of products than there would be no sense in extending the area of land under cultivation. It is also suggested that additional quantities of grain production would be possible on the same piece of land.

Later on, with the generalization of real world situation of land use, Von Thunen formulated a ring of varying intensity of land use and cropping pattern around settlements. The modifying impact of transport route is also shown on the same model. Further, this author has devised polar and anti-polar concepts of land use. The polar concept of land use is based on the assumption of idealized polar condition of the earth and this concept is mainly used here to undertsand the world rural land use in relation to a human habitation.

In land use study the consideration of the centre of the village as a pole is not appropriate because on poles the intensity of land use is low and it is increasing towards equator, while near the village the intensity of land use is high and away from the village it is low. Thus rural land use is rightly known as anti-polar concept of land use in relation to a rural settlement.

Technique of land use survey and mapping includes the materials required for the survey, field investigation for relevant data and land use, their systematic arrangement under different sub-heads, quantitative representation of various data by different formula and cartographic arrangement of these materials. In this study only elementary techniques, sampling designs and statistical methods have been given so that it is easy to study land use with greater enthusiasm on lives of modern quantitative revolution.

Land use classification has received great attention of geographers in different parts of the world. It is in common parlance in any study to classify the land use in different categories. The land use classification of the world, the U.K., the U.S.A., the U.S.S.R., Japan, Poland and various other classifications related to Indian situation are delineated under this aspect. Ultimately a case study of Bihar has been presented here which is closely related to the other Indian studies. This shows that land use classification depends upon the climatic and terrain conditions to the region and the needs of the society.

The term 'land capability' is broad and relative which shows that certain land can produce better crops with fewer conservation practices than others. This classification suggests suitability of different soil types for production of better crops in different ecological conditions and in different regions. In this study positive and negative variable which broadly control the land capability have been determined. In any region the desirable amount of rainfall, plain land, soil texture and types, and different modes of irrigation act as positive factors of raising land capability, while the ruggedness of land, flood, drought, forest cover, water logging and soil erosion act as negative variables of land capability. After weighing the positive and negative variables in shaping the land capability, different regions of land capability have been determined for the conservation of land and planned agricultural development. Afterwards, the dispersion of land capability has been tested in relation to the net sown area, and also on Lorenz Curve. At the end, the land capability index required for certain land use is also determined.

The study of land capability has also a merit to help in agricultural planning and providing guidance for the better use of the land in a particular capability class zone. Hence, it is expected what has been given here may sound as an innovation in the field of land use, agricultural geography and planning.

The variation in agricultural efficiency in different districts is mainly due to various topographic features, fertility of soil, intensity and amount of rainfall, irrigation facilities and development of transport routes in Bihar. Generally the northern plains are producing better crops than the plateau regions of Chota Nagpur, because the former have fertile soil in comparison with the latter.

The section on crop combination deals with several methods and finally a new method has been designed here which can be applied universally. This new method of crop combination has been formulated clearly with decreasing intensity of crop land use and its regional signficance, so that the ills of low percentage of crops can be removed by technological advancement, fertilization, conservation of soil, controlling weeds, pests and flood menace, and with the provision for irrigation and better seeds.

In land use study, land resource is generally measured in terms of productive capacity and environmental resistance. Food production is directly or indirectly controlled by the land and its conversion into calories gives a scientific measure of land resource as it is necessary for sustenance of human health. Calories intake by an individual is also measured here in terms of 'standard nutrition unit' (SNU) and potential productive unit (PPU). The average SNU of 800,000 calories as fixed for the average Indian adult, while SNU produced by foodgrains produce is only 500,000 calories, due to the lesser production of foodgrains in comparison with the population pressure in Bihar. The balance is compensated by other means. Thus the further increase in population requires more efficient use of the land. This requires the adoption of scientific way of cultivation besides raising considerable amount of fruits, vegetables, fishes and other items of our daily food.

The preservation of natural resources of the land e.g. soil, vegetation, water and animals from destruction by protective management for the benefit of the coming generations is known as conservation of land resources. In this regard, the misuse of land means the loss of resources conservation and, the judicious use means greater good of the society.

Land reform is related to the proposals of acquiring surplus land and its redistribution among the poor landless or the better utilization of the available land. In India, high population pressure requires urgent attention. But the main factors for tardy progress are the state governments themselves who have failed so far in generating political will, firm determination, better land policy and in mobilising public opinion to sustain land reform measures. It is necessary in order to remove poverty in rural areas. This should be judged on the basis of social-justice and economic conditions of people prevailing in India.

So far as the planning proposals are concerned, the peasants should leave their traditional land use practice in respect of changing man-land-ratio and scientific development. The government and peasants should provide irrigation to each and every field. They should practise cooperative farming for the integrated area development and better training facilities in bigger cooperative farms having good transport network and communication lines in rural areas for the movement of machines and men. In some areas these movements have already been followed by the peasants under crash programme and 'grow more food' campaign. Thus, under such a changing land use practice, it is quite essential to note down the traditional pattern of land use and remove the old system of cultivation for feeding the millions of growing people of our villages and towns so as to rise economic status and to acquire self-sufficiency. Only the planned survey and resource assessment can really help the peasants in land use planning.

Index